Thinking about Physics

Thinking about Physics

Roger G. Newton

PRINCETON UNIVERSITY PRESS

PRINCETON AND OXFORD

Copyright © 2000 by Princeton University Press
Published by Princeton University Press, 41 William Street, Princeton,
New Jersey 08540
In the United Kingdom: Princeton University Press,
3 Market Place, Woodstock, Oxfordshire OX20 1SY

The Library of Congress has cataloged the cloth edition of this book as follows

Newton, Roger G.
Thinking about physics / Roger G. Newton.
p. cm.
Includes bibliographical references and index.
ISBN 0-691-00920-1 (cl : alk. paper)
1. Physics—Philosophy. I. Title.
QC6.N485 2000
530′.01—dc21   99-35807

British Library Cataloging-in-Publication Data is available

This book has been composed in Palatino

Printed on acid-free paper.

www.pup.princeton.edu

Printed in the United States of America

2 3 4 5 6 7 8 9 10

TO THE MEMORY OF

JULIAN SCHWINGER

Contents

Preface

THIS BOOK, addressed to readers with a good undergraduate education in physics, is about some ideas that are usually not discussed in college courses, at least not explicitly, but which are nevertheless an essential part of our science. University teaching is generally very much oriented toward problem solving, and some physicists, unfortunately, assume that is all there is to their discipline. No doubt problem solving, in the widest sense, is our bread and butter; but it is surely not all we do in our professional lives. The way we approach obstacles in research is influenced by our more general understanding of what lies behind the solutions to large problems tackled in the past, and we cannot help reflecting upon the deeper and more general issues, spending some time *thinking about physics* rather than simply *doing it*.

While routine and most not-so-routine physics problems have unique solutions which, once found, can produce no reasonable disagreement, it is in the nature of some of the issues we will be discussing that they are subject to strong disputes among intelligent people. Perhaps these issues will seem, then, to belong to philosophy, a word I dare only whisper, because I know that many of us shy away from it. After all, branding an idea *metaphysical* is to use about the most dismissive term in our vocabulary, but of course we are all subject to at least a small dose of metaphysical preconditioning, aren't we?

As a corollary of the fact that some of the questions I raise are very subtle and the answers subject to debate, I advise you not to accept my word without question, but to use my arguments as starting points for your own thinking. I am not going to lay out all the solutions that have been offered over the years, with pros and cons, but I will offer my own for you to consider.

In my presentation, I want to demystify quantum mechanics as much as possible. That, of course, cannot be completely accomplished, but I would like to separate those aspects of quantum physics—and there are plenty of them—that may appear strange yet are intrinsic to any probabilistic rather than strictly causal theory from those that are *weird* in the much more profound sense characteristic of the quantum theory. Therefore, whenever I point out that a certain counterintuitive feature is "no more than what can be expected in any probabilistic theory," you should not conclude that is all there is to the strangeness of the quantum world. I will discuss primarily in the last chapter the specific aspects of this world beyond the consequences of ordinary use of probability.

Another feature of this book is my disagreement with a number of prominent physicists, such as Feynman and Heisenberg, who, at the most fundamental level, give primacy to the particle concept. My attitude, by contrast, is to regard the *quantum field* as the basic entity, with "particles"—possessing great intuitive appeal but circumscribed utility—appearing as phenomena produced by the field. Many of the quantum paradoxes, I will argue, have a linguistic source, stemming from the use of the concepts of particles and waves, to which our everyday intuition and language seem to drive us, but the connotations of which, originating from the macroworld, are simply inappropriate to the microworld.

This book, however, is not limited to quantum mechanics. We will also be discussing aspects of classical mechanics, field theory, and thermodynamics as well as the role of mathematics in physics, the concepts of probability and causality, and the various arrows of time and their relation to one another, in short, all the basic notions we usually take for granted when doing physics and rarely mention explictly.

At this point I want to acknowledge my debts. There are, of course, too many colleagues and friends to mention from whom I have learned much over the years, and whose wisdom can be found in this book, but the three I want to name particularly are, first of all Julian Schwinger, whose teaching had a profound influence, and later Ciprian Foias and Larry Schulman. I also owe a large debt of gratitude to my wife, Ruth, for untiring, invaluable editorial assistance.

Thinking about Physics

Introduction

WE PHYSICISTS are rightly proud that ours is an "experimental science." Does this mean that all we do is tinker with bits of sophisticated apparatus and machinery, trying our best by trial and error to find out how nature works? Surely physics is more than a collection of experimental results, assembled to satisfy the curiosity of appreciative experts or applied to serve the needs of humanity.

When we use the term "experimental science," we want to distinguish ourselves from philosophers, especially so since in former times, "natural philosophy" was the name for what we do. To call physics *experimental* is to emphasize, as Isaac Newton did in his *hypotheses non fingo* that our theories are not fanciful speculation but are solidly grounded in experimentally verifiable facts, although they are capable of soaring as high as any poetic imagery and delving as deep as any philosophical thought. We side with Robert Boyle, the originator, after Galileo, of the modern experimental method in science, who in controversies was able to point to the data he had obtained by means of his refined air pump, rather than with his adversary Thomas Hobbes, for whom the results of laboratory work were philosophically irrelevant. Hobbes deemed metaphysical such concepts as the vacuum, to be examined not by dirty experimentation but by pure thought alone. In contrast to other intellectual debates, which may swirl forever without being settled, conflicts between rival physical theories can be resolved by appeal to observation and experiment, the results of which are open to public inspection. Though we are often accused, perhaps with some justification, of being overly aggressive, there have been, for a long time now,[1] very few *ad hominem* attacks or harsh criticism of the work of others in our literature.

---

[1] This was not always so. Here is Isaac Newton's reply to some criticism by Robert Hooke of his 1672 paper in the *Philosophical Transactions* of the Royal Society on the nature of color: "Mr Hook thinks himselfe concerned to reprehend me for laying aside the thoughts of improving Optiques by *Refractions*. But he knows well y^t it is not for one man to prescribe Rules to ye studies of another, especially not

We do not, however, collect observational or experimental facts for their own sake. Physics left the stage of taxonomy and classification of data more than two thousand years ago. The mere minute description and naming of curious and noteworthy effects or entities, the dominant goal of younger sciences, is of little interest to most physicists. One of Richard Feynman's seminal recollections from his childhood was his father's insistence that just knowing the name of a phenomenon does not help us understand it. What we are generally looking for is *explanations*. Even though Otto Hahn received the Nobel Prize for the surprising detection of barium when neutrons collided with a uranium target, his experiment with Strassmann would have had little impact without the subsequent theoretical interpretation by Otto Frisch and Lise Meitner; it was only on the basis of the work of the latter two that Hahn and Strassmann were credited with the discovery of nuclear fission.[2] Finding unfamiliar spectral lines in the light of stars was no more than a puzzling curiosity until Niels Bohr, using his new quantum theory of atomic spectra, was able to explain quantitatively that the light must have been emitted by highly ionized atoms. This explanation served as such a persuasive confirmation of his revolutionary quantum model that it was reported to have convinced even the skeptical Einstein, who frowned on Bohr's acausal quantum jumps. Experimental facts are of central importance for physics, but only in the context of a theory—the theory leads us to understand the facts, and the facts, in turn, undergird the theory.

Clearly, then, the role of experimenters in physics is crucial. They serve two equally important functions: exploration and testing. By exploring new areas of nature to find significant new facts or probing the reactions of familiar objects to unfamiliar conditions, the experimentalist directly extends our knowledge and pushes for-

---

without understanding the grounds on wch he proceeds." Quoted in Richard S. Westfall, *Never at Rest* (Cambridge, U.K: Cambridge University Press, 1980), pp. 246–247.

[2] Regrettably, Frisch and Meitner received very little of the credit and did not share in the Nobel Prize.

ward the scientific frontier, opening up new fields and finding novel phenomena that need to be understood. The greatest discoveries are those of interesting new facts as yet unexplained, and few such findings are serendipitous. Equally important is the testing role. It is up to the person in the laboratory or in the observatory to submit any theory, put forward for the purpose of explaining other facts, to the most rigorous experimental or observational examination, in an attempt either to verify predictions implied by the theory or to falsify it. While confirmation of a prediction may lead to a significant discovery and to fame, the shooting down of a theory is less glamorous; generally, however, there is more leeway in a verification, and unavoidable biases tend to work in favor of it; falsifying is usually more clear-cut and less ambiguous (with caveats to be discussed below).

Every experimenter also has to confront the question of whether to test a well-established theory or a daringly novel one. Corroborating an accepted theory or demolishing a new one may be easy, and the rewards accordingly meager; attempting to prove that a theory of long standing has to be modified, or providing support for a startlingly fresh one, may be a difficult long shot, but that's where the glory waits. The choice is that of a gambler, and there can be no general rules.

THEORY-DEPENDENT FACTS

The symbiotic relationship between experimental data and theoretical structures has an important consequence that is sometimes misinterpreted by nonscientists: to some extent the very existence of facts, or at least their formulation, depends on the theory in which they are embedded or for whose support they serve—many facts are "theory laden." For one thing, we employ theories, usually well-established ones, to discover new facts. Think of the extent to which Robert Millikan relied on the known laws of electromagnetism and viscosity to measure the charge of the electron, and how astronomers depend on the constancy of the speed of light (and therefore the special theory of relativity) to measure, or even to state, distances to stellar objects.

Astrophysics, cosmology, and cosmogony are the branches of physics in which our factual knowledge is, on one hand, necessarily gained primarily by passive observation[3] rather than active experimentation, and, on the other hand, is most heavily indebted to theory. Distance measurements to galaxies rely fundamentally on observations of Cepheid variables: the intrinsic luminosities of these stars are, on the basis of detailed, well-understood models, theoretically expected to be correlated in a known manner with their periodicities, so that their apparent luminosities allow us to determine their distance by the inverse-square law. The recession velocities of stars and galaxies are inferred indirectly from their redshifts. Therefore, Hubble's law, though it appears to be based almost directly on observation, cannot be divorced from a heavy reliance on theory. The farther back you go in the history of the universe, the more necessary are the crutches of theory-based inferences to establish what we regard as the facts of cosmogony.

In the areas of physics dealing with the very small, where experimentation rather than patient observation is the dominant mode, we also have to depend heavily upon theory for ascertaining facts. How did Ernest Rutherford discover the nucleus? He inferred the existence of a central heavy charge from the backward scattering of alpha particles, which classical electromagnetic theory told him would never happen if the positive charge in the atom were distributed over its entire volume, as J. J. Thomson's model had it. (He was remarkably lucky that the classical scattering cross section in this special case happens to coincide with the one predicted by quantum mechanics, which in 1911 was still unknown!) Ever since Rutherford, all we know about the structure of the nucleus we have learned by means of scattering experiments, whose interpretation would be impossible without quantum mechanics.

---

[3] The word *passive* here should not be misinterpreted to mean that the observer simply sits back and does nothing but look. Observation is a highly selective process, in which the activity of the observer consists in the precise direction of her attention: looking is not the same as seeing.

We cannot avoid the use of theoretical insights in the search for new facts. Indeed, using the most sophisticated understanding gained by theory in the design of the apparatus and in the strategy of any search for new phenomena is the mark of an ingenious experimenter. There are few areas at the forefront of physics today where experimental facts can be ascertained by direct apprehension. Michael Faraday lived and worked at a time in which physics was at a simpler stage of development and long strings of theoretical deductions were seldom necessary in order to dig out facts; that is why—apart from his cleverness and the circumstance that the gigantic pieces of apparatus often needed at present were not yet required—he was able to perform his public demonstration lectures about his latest research so successfully.

Facts sometimes are theory laden in another sense as well: their very meaning may depend on the context of a theory. What a high-energy physicist means by announcing the discovery of a new unstable particle is usually no more than the observation of a noticeable bump in the plot of a scattering cross section as a function of the energy, with the center of the bump indicating the new particle's mass, and the inverse of its width its lifetime. (We shall return to this point in the last chapter.) For many years, the significance of the measured redshifts of stars was considered controversial: in the context of the theory of relativity, such shifts meant that the star was receding, while in the context of other theories or conjectures they indicated no such thing.[4] In addition, the very recognition of a redshift depends on the identification of light as being emitted by an element whose unshifted spectrum we know on Earth.

Once we realize that many of the facts on which the structure of physics rests are theory laden, it becomes natural to ask, what justifies our elaborate theories? After all, if scientific theories are distinguished from philosophical speculation and primitive myths by virtue of their being ultimately grounded in facts, and if the theories themselves to some extent determine these facts, are we not caught in a vicious circle? This is precisely the entry point

---

[4]The redshifts were conjectured by some astrophysicists to originate simply from their travel through space and a resulting loss of energy of the photons.

7

for some of the critics of science who claim that we are simply a group of ingrown self-declared experts whose understanding of nature has little cognitive value.[5] Before I answer to counter their criticism, let us look a bit at theories and their purposes.

Why do we need theories? Some contend that their only purpose is to provide an economical description of the facts: we spell out laws of nature in order to provide a simple summary of masses of experimental data—just as the formula $x_n = n^2$, where n are the natural numbers, encapsulates the infinite sequence 1, 4, 9, 16, ... Such a view of theories in physics is surely inadequate. Of course, theories do serve the purpose of describing the behavior and the features of nature by very economical means—the more parsimonious in their assumptions, the simpler they are, the more we admire them. But empirical laws, such as Ohm's or Boyle's, do that too, and this is their only aim. An economical description of a mass of data is certainly not the only function of what we dignify with the word "theory"; their principal purpose is not merely to summarize the facts but to *explain* them. In chapter 1 we shall discuss this point in more detail.

WHAT NEEDS EXPLAINING?

It is one thing to debate and discuss the modes and the nature of explanations, but there is also the deeper question, what needs an explanation? The answer has varied considerably in the course of history. There are times when physicists ask certain questions and search for answers, other times when these same questions remain unasked because they are considered uninteresting or irrelevant. When Isaac Newton promulgated his law of universal gravitation, he did not ask by what mechanism the gravitational force was transmitted through space, even though his own postulated "action at a distance" made him, as well as his contemporaries, at first extremely uncomfortable. René Descartes had formulated

[5] For more details, see my book, *The Truth of Science* (Cambridge, Mass.: Harvard University Press, 1997).

8

an intricate model in terms of vortices, but for Newton no more detailed explanation than the mathematical expression of his law was needed. Even though we are today again equally uneasy with action at a distance—Einstein's general relativity replaced it by geometry or, in effect, by a field theory—ignoring the need for an explanation turned out to be enormously fruitful.

In the same context, there were other features of planetary motion—the fact that all the planets revolve around the Sun in the same plane and in the same sense—which Descartes tried to explain but which were ignored by Newton. Here the absence of an explanation is a profound matter of principle. The Newtonian laws of mechanics are differential equations, whose solutions require initial conditions, and these are furnished by the contingencies of nature rather than by the laws. In contrast to earlier times, when elaborate models with Platonic ratios were invented to account for the distances of the planets from the Sun, Newton's laws allowed the planets' distances, the coplanarity of their orbits, and their sense of motion to be accounted for by the accidents of their formation. As far as the new physics was concerned, they needed no explaining.

We have here perhaps the most significant characteristic of physics introduced by Newton. The laws by themselves are not designed to explain the course of the universe or all the features of the solar system. The *laws* are meant to be universally applicable, and in order to explain the motions of a specific system of objects they have to be supplemented by initial conditions, whose origins are to be sought elsewhere. Of course, the beginnings of the solar system can also be understood by means of physics, but the dynamics of its formation are separate from the dynamics of its steady state. Similarly, physics does not uniquely explain the history of the universe; it determines that history only once the contingencies of the big bang are postulated.[6]

---

[6] According to recent speculative ideas, the equations of some cosmological models generate their own boundary conditions at the initial singularity. If that turns out to be correct, my statement will be invalidated.

Ever since Newton, we have expected cosmogony to be not simply history, but a history derivable from, and therefore explained by, universal laws (plus initial conditions)—laws that themselves do not evolve in time. However, Paul Dirac[7] and others have speculated that the fundamental constants of nature depend on the size of the universe and therefore are changing in the course of time. There is at this point no evidence for such change—in fact, much evidence argues against it[8]—but if this conjecture should turn out to be correct, the very time dependence of the laws would require an explanation. In other words, we would have to search for an underlying *time-independent* law that would account for the specific way in which these constants vary with time. Physics never regards history itself as a sufficient explanation of any fundamental change.

Quantum field theory represents an example of a different kind of shift in what we expect a theory to explain. The marvelously intricate theory of quantum electrodynamics, which has been enormously successful in accounting for many detailed effects of the interaction of electrons with photons, provides no explanation for the numerical values of the mass and charge of the electron. Moreover, even if these numbers are inserted "by hand" in the equations, quantum electrodynamics (QED) predicts that they will differ from their observed values because of the interaction of the electrons with the electromagnetic field, and these *renormalized* values of the charge and mass are not calculable either—they come out to be infinite. The success of the renormalization program that Richard Feynman, Julian Schwinger, and Shinichiro Tomonaga initiated was based on using the experimentally observed mass and charge in the theory as their renormalized values. (The unrenormalized values appearing in the equations were then never needed and remained redundant; they could not, however, be assumed to

[7] P.A.M. Dirac, "The cosmological constants," *Nature* **139** (February 20, 1937), pp. 323–324.

[8] For example, the structure of the spectra of far-away galaxies, whose light was emitted long ago, is the same as that of elements at the present time (apart from the redshift) and shows no evidence of different values of the mass or charge of the electron at the time of emission.

be zero, tempting as such an assumption may have been.) The success of this theory in predicting other numbers with great accuracy therefore rested in part on a conscious renunciation of the need to explain certain observations, and, though some prominent physicists—including Dirac—were quite unhappy with this procedure, most of those working in the field were satisfied.

Recent theories of elementary particles have been more ambitious, though, and while none of them is yet in a position to match the predictive power of QED, they are expected to explain quantitatively the masses of all particles (except possibly for one, which would set the scale) and their interaction strengths. The aim of those working on a "theory of everything" is to account for the behavior of the universe and the properties of its constituents without inserting anything "by hand." Some even go so far as to regard the question "Why existence?" as the most important in physics.[9] To my mind, this is a metaphysical question if there ever was one, which can never be answered by physics; one may well ask if it has any scientific meaning at all.

WHAT SERVES AS AN EXPLANATION?

Clearly, the meaning of the word *explanation* depends at least in part on the preparation and expectation of the listener. In Greek mythology, the Earth was supported by Hercules, in other myths, by an elephant, but we do not accept these as explanations of what "holds the Earth up" in space. While mythological answers often depend on models whose pictorial power is sufficient to capture and overawe a listener's imagination, we require either wide-ranging principles or mechanisms relying on causality. Even for physicists, however, there are large differences among the needs that a satisfactory explanation has to meet.

We all know there are explanations of a phenomenon some physicists might regard as satisfactory but others—not to speak

[9] John A. Wheeler, *Geons, Black Holes and Quantum Foam* (New York: W. W. Norton, 1998).

of nonscientists—may consider insufficient, particularly when it comes to the role of abstractions. When a theorist offers a detailed mathematical explication of an effect, many less mathematically inclined experimenters may accept the underlying theory and yet deny that they have been led to any real understanding. The difference lies mainly in the appeal to and use of their intuition. Just as many physicists working with experimental equipment have a very tactile imagination about the objects under investigation, mathematically oriented theorists often develop a strong mathematical intuition; the former may ask for a concrete physical agent, while the latter are quite happy with a "mathematical mechanism."

Take, for a very specific example, the behavior of the cross section for scattering of a particle by a bound system, like an atom, in its ground state near the threshold energy for lifting the atom to an excited state: as a function of the energy, the curve will exhibit a *cusp* with infinite slope. There is a relatively simple way of explaining this phenomenon purely in mathematical terms by showing that the element of the S-matrix[10] corresponding to the inelastic reaction is proportional to the square root of the momentum of the outgoing fragments (so that the corresponding cross section is proportional to their velocity), and since this momentum vanishes at the threshold, its derivative with respect to the projectile energy—a linear function of the square of the momentum—must be infinite there. It then follows from the unitarity of the S-matrix that the element for elastic scattering must also have an infinite derivative.

Those who are not mathematical physicists, however, will regard such an analysis as quite unsatisfactory and will be much happier with the following way of explaining it: since the inelastic cross section is proportional to the outgoing flux, it has to be pro-

---

[10] The S-matrix relates the behavior of a quantum system in the far future, when it will be observed, to that in the distant past, when it was prepared. Heisenberg regarded it as the most fundamental theoretical tool in physics and attempted to replace the Hamiltonian with it, a program that never succeeded and is now out of fashion. Nevertheless, the S-matrix is an enormously useful instrument for linking basic theory to observational data.

portional to the momentum of the excited system and hence must, as a function of the projectile energy, start with infinite slope at the threshold as the incoming energy is raised. But since the total flux is conserved, the sudden removal of flux (with infinite slope, as a function of the gradually changing energy) by the opening of the new inelastic channel leaves less flux available to the elastic channel; this reduction must therefore also happen with infinite slope, producing a cusp in the elastic cross section. Even though this second description of the effect is nothing but a translation of the first into more physical language, it has greater explanatory power for most physicists.[11]

In a more general vein, Einstein made an important distinction between "constructive theories," such as the kinetic theory of gases, on one hand, and "theories of principle," such as thermodynamics and relativity, on the other; the former offer detailed mechanisms while the latter present abstract principles of wide scope and great "heuristic value," by which he meant that they had particular "significance for the further development of physics."[12] Only the first, he asserted, led to real understanding: "When we say that we understand a group of natural phenomena, we mean that we have found a constructive theory which embraces them."[13] No grand symmetry principle, for example, however useful and powerful, can, in Einstein's view, ever lead to real *understanding*. While his distinction between theories is important, and the purposes the two kinds of theories serve are different, I am inclined to disagree with Einstein. The theories of principle, I believe, serve as explanations as well. When we derive the Maxwell equations from general principles of relativistic invariance and simplicity, rather than from detailed mechanical models as Maxwell did, we experience the same feeling of satisfaction we obtain from any mechanistic explanation. And should we ever arrive at a *theory of everything*

[11] Both kinds of explanations are given in more detail in my book, *Scattering Theory of Waves and Particles*, 2nd ed. (New York: Springer-Verlag, 1982), pp. 538ff.

[12] See Martin Klein, "Some turns of phrase in Einstein's early papers," in *Physics as Natural Philosophy*, ed. A. Shimony and H. Feshbach, pp. 369–373 (Cambridge, Mass.: MIT Press, 1982).

[13] A. Einstein, *Out of My Later Years* (Secaucus, N.J.: Citadel, 1956), p. 54.

13

as the only mathematically consistent scheme underlying the behavior of the universe—unlikely as I think it is that this will ever come to pass—we would surely count it as an explanation of the way things are.

We simply have to stretch the meaning of the words "understand" and "explain" to go beyond mechanisms and to include general principles. When Isaac Newton introduced his law of gravitation to explain the motion of the Earth and the planets as well as that of the falling apple, he was attacked precisely because he did not provide a mechanistic construction of the force of gravity, as Descartes's theory of vortices had done. Yet we feel today that his laws have led us to understand planetary motions of the solar system, despite the fact that some nonphysicists, asking *why* gravity follows an inverse-square law, conclude that Newton's laws do not explain anything.

When we understand something, we can fit it comfortably into our previous body of knowledge and feel that it harmonizes with our intuition. Mastering an argument is more than being able to reproduce it. When we see a mathematical theorem and its proof for the first time, we may be able to memorize and restate it verbatim (as students are required to do in some schools) without understanding, but there comes a time when we experience that mental click, the "aha!" when we suddenly feel we get it. Everyone working in mathematics has had this experience; a student commonly agrees with every step in a long and complicated proof and yet feels he does not *understand* it. And a physicist may see a natural phenomenon, possibly may even be able to fit it roughly into some theoretical framework, but might still feel she does not understand it. The sudden click of comprehension will clearly depend on the makeup of the listener's mental preparation and intuition.

ANALOGIES

One of the most useful tools for setting off the click of understanding is to establish an analogy between a new phenomenon and one already understood. We are all accustomed, for instance, to certain parallels drawn between resonances in atomic or nuclear physics and their counterparts in mechanical systems or electromagnetic circuits. In the early days of quantum mechanics, the fact that the new theory had certain features in common with counterparts in the behavior of electromagnetic fields helped some physicists understand it more easily. For many years, a grounding in electromagnetism helped students grasp certain aspects of quantum mechanics more easily as well. Today, however, when it is customary for them to learn quantum mechanics before they get to the more intricate properties of classical electromagnetic waves, I have found that it advances their understanding of the details of the angular distributions of multipole radiation to draw analogies—and that is all they are at the classical level—to quantum mechanical angular momentum, with which they are already familiar. Has anyone ever learned about the redshift of light from distant galaxies without having it explained via the Doppler shift in acoustics and the sound of a moving siren?

Useful as analogies are for understanding, they can also be dangerous. In the first place, they should be employed only as helpful crutches or auxiliary tools that can later be discarded. As such they are also extremely useful in stimulating the imagination to find a new theory. On the other hand, if analogies are used *in place* of a real theory, as they sometimes are in younger or less developed sciences, they may lead only to the *illusion* of deeper understanding. A serious scientific explanation cannot rest with an analogy or a metaphor; such devices must never be more than stepping-stones or supplements to the real thing, based on theory.

That they may be misleading is another problem in the use of analogies. A prominent instance of this is the transfer of Heisenberg's indeterminacy relation to other sciences. In many contexts far from physics, such as economics and psychology, for example, one sometimes finds statements to the effect that, "as Heisenberg has shown," when we measure a quantity, we interfere and disturb the system under investigation in an uncontrollable manner. This true but rather trivial assertion is based on an erroneous interpretation of the basis of Heisenberg's relation, which ultimately

15

goes back to a misunderstanding of Heisenberg's original paper[14] with its discussion of the "gamma-ray microscope," a hypothetical experiment demonstrating that the precise localization of an electron requires light of short wavelength—therefore a high-energy photon—which gives a kick to the localized electron and alters its momentum. In the context of the Einstein, Podolsky, Rosen (EPR) debate in chapter 8 it will become clear that this experimental-disturbance origin of the indeterminacy relation, intuitive though it may be, cannot be seriously maintained. Therefore the transfer, by analogy, of the Heisenberg relation to other contexts is usually seriously misleading and designed only either to give artificial support to reasoning that may be strong enough to stand on its own, or to bolster arguments too weak without the purported crutch from physics.

When we think about what may serve as a satisfactory explanation we move on to a closely connected question about the *concepts* introduced by a theory and how theorists use such ideas as tools. How do we define an electric field or friction? As physicists we are usually careful to introduce such concepts in experimental terms: their meaning is derived from the methods that allow them to be measured by scientific instruments. Emphasizing the need for experimental definitions of all concepts used in science, a group of physicists—the school of instrumentalism—at one time refused to accept any theoretical ideas that could not be directly observationally verified. A theory they found particularly objectionable from this point of view was Einstein's general theory of relativity with its reliance on such abstract notions as the metric tensor and curvature of space. This school has pretty much faded away now, and most physicists realize that such a severe restriction on the admissible conceptual tools of a theory—to admit only those amenable to direct observation—is far too limiting. While all physical theories, in order to have meaning, must lead to consequences that can be verified by scientific instruments, either experimentally

---

[14] Werner Heisenberg, "Über den anschaulichen Inhalt der quantentheoretischen Kinematik und Mechanik," *Zeitschrift für Physik* **43**, 172 (1927).

or observationally, there is no reason why every auxiliary concept used in them should have an experimental meaning. Nevertheless, there is no denying that when some theorists go too far in their use of abstract ideas and let their mathematical imagination soar too high above the observational ground, it does make others, especially experimenters, uneasy and prevents them from accepting such theories as legitimate explanations.

The Anthropic Principle

A number of physicists have adopted a mode of explaining the numerical values of the fundamental constants of nature that is radically different from the usual one. The basis of their argument is the following set of facts. If the gravitational constant were larger than it is, the initial outward momentum of matter after the big bang would not have been able to overcome the gravitational pull, and the universe would have fallen back into itself before becoming more or less isotropic as we find it now on a large scale. If that constant were smaller, the explosion would have propelled matter out, and no galaxies and star clusters would have been able to form unless there had been strong inital inhomogeneities, which in turn would have led to large-scale anisotropies in the present universe and in the remnant of the primordial black-body radiation; for such anisotropies there is no evidence. Without stars, life would be impossible. Furthermore, if the strong coupling constant were smaller than it is, the nuclear force would be so weak that no nuclei heavier than hydrogen could be formed, and, again, no life could exist.

The *anthropic principle* contends, remarkably, that these facts can serve as "explanations" of the values of these constants, because without life, there would be no intelligence and humans would not exist. "Since it would seem that the existence of galaxies is a necessary condition for the development of intelligent life, the answer to the question 'why is the universe isotropic?' is 'because

17

we are here,' " assert C. B. Collins and Stephen Hawking.[15] There are several different ways of interpreting this "because."

We need not concern ourselves with a possible religious interpretation, that human life represents a *purpose* underlying the structure of the universe, or with a conceivable philosophical one in the sense of Bishop Berkeley's idealism, that without a human observer the universe would not exist, though some physicists may perhaps subscribe to this. One possible meaning of the anthropic principle in terms of physics might be constructed on the basis of the "many worlds" interpretation of quantum mechanics.[16] According to this interpretation the quantum mechanical probabilities arise from the fact that the world is continually splitting into infinitely many separate realities, each of which incorporates one of the possible outcomes of a measurement. We could then say there are infinitely many universes with different values of their coupling constants, and naturally we would have to inhabit one of those in which life can exist. Indeed, this construal might be imagined even without the "many worlds" interpretation of quantum mechanics, especially in view of some recent speculations that parallel universes are constantly being born at the end of umbilical cords generated in black-hole singularities.

No matter how we construe the anthropic principle, whose essence is the reasonable contention that *it is no accident that we find ourselves in a world in which intelligent life can exist,* if used as an explanation of certain details of the universe, it reverses what is usually meant by the word "explanation."[17] It is as though the

---

[15] C. B. Collins and S. W. Hawking, "Why is the universe isotropic?" *Astrophysical Journal* **180** (1973), pp. 317–334.

[16] H. Everett III," 'Relative state' formulation of quantum mechanics," *Reviews of Modern Physics* **29** (1957), pp. 454–462; see also B. DeWitt and N. Graham, eds., *The Many-Worlds Interpretation of Quantum Mechanics* (Princeton: Princeton University Press, 1973).

[17] The same argument applies to the idea that our presence serves to explain why we live in a universe with as much order as is necessary for our existence. Large fluctuations will surely occur in the course of the development of the universe in various regions, and it is no coincidence that we find ourselves in a region orderly enough for us to live in. But that cannot be regarded as an explanation of this order.

existence of humans were taken to explain the extinction of the dinosaurs; after all, it was their extinction that made the evolution of mammals possible; intelligent life would not have emerged without it. Would anyone take such an argument seriously? The only interpretation of the anthropic principle that would seem to me to be acceptable is one in which it is not used as an explanation of the size of the constants of nature: if these constants indeed depend on the size of the universe and hence vary with time, it is certainly not surprising that we live at a time when their values allow life to exist. In this construal, however, the dependence of the cosmological constants on the size of the universe needs explaining in terms of a time-independent theory rather than in terms of our existence.

COHERENCE OF THE WHOLE

Returning to the fact that the tools for our explanations are elaborate theories, which in turn are ultimately grounded on experiential data, we have to acknowledge—denigrating neither the experimental facts nor the theories they corroborate—that the notion of free-standing experimental discoveries of pristine facts of nature constituting the foundation on which the theoretical edifice of physics is erected is far too simple. What justifies our confidence in the basic soundness of the entire structure is its *coherence*, an intellectual coherence that includes consistency with all the experiences and expectations founded on it, the fulfillment of precise, far-reaching predictions implied by it, and the functioning of all the technology built on its basis. To point out that *science works*, in the sense that we readily watch television and have confidence in the airworthiness of the next plane we take, is, of course, the most banal of the answers we can give to those who question the truth of science; it is, nevertheless, an important component of the coherence of physics, which is why we can legitimately claim science *does* work. It is this coherence that gives the scientific enterprise its stability, just as the stability of a high-rise building depends on the ubiquitous cross-bracing in its walls and floors.

While the discovery of some new facts may lead to the falsification of a specific theory, it cannot make the entire structure of physics crumble—it would take a very large hole in the side of a well-constructed skyscraper to make the building collapse. Controversial as it was even a hundred years ago, the molecular structure of matter, for one example, could not conceivably now be found in error; too many pieces of evidence from too many different kinds of experiments point to and corroborate it. Coherence, understood in the broad sense outlined and not just in the sense of internal consistency, is therefore what, in my view, is the basis for claiming *truth* as the goal of physics—not as an attainment, but as an aim.

The coherence of the structure of science makes it often possible for us to reject, without detailed examination, claims made in support of extraordinary effects, such as those in parapsychology. No doubt, these rejections may appear biased or prejudiced—literally prejudging reports on the basis of what we know about how the world in general works—and on some rare occasions the closing of our minds to purported new phenomena may turn out to have been ill-advised. But such novelties, if legitimate, will eventually make their way into science and will be accepted. We do not disbelieve in precognition or psychokinesis simply because parapsychologists lack the proper credentials or organizational clout to make themselves heard, as some commentators allege; we recognize their incongruity with all we have come to know about nature. This century has had more than one "paradigm shift" in physics, and few physicists would believe we have seen the last. But the fact that revolutions have indeed occurred and were of great benefit to physics does not mean that every wild proposal is a great potential advance. To appear crazy at first may be a necessary but certainly is not a sufficient condition for a revolutionary new insight. By and large, science is a relatively conservative enterprise, yet like a good pressure cooker with a reliable safety valve, it can accommodate itself, when necessary, to its internal revolutionary tensions without being destroyed.

The central place occupied by coherence makes it the quintessence of what we mean by *truth*. While it would be a mistake to transfer the meaning of this word in science to other contexts—as in the esthetic, religious, emotional, or philosophical senses—truth does play an important role in the work of physicists, Thomas Kuhn's denial and our own sometimes cynical front notwithstanding. We have learned enough about the strangeness of nature not to be taken in by any pretensions that what we know and say about the world *corresponds* in some literal and mirror-like sense to what is out there. Since such an idea could not survive confrontation with the quantum theory, some physicists have retreated to a position of regarding all theories as nothing but *models,* thereby seeking to sidestep the question of truth altogether. But this apparently safe position is surely not shared by many experimenters, who, feeling themselves in touch with nature most directly, deem submicroscopic particles to be as palpable as they were for Rutherford. (We shall turn to the question of reality from the perspective of quantum mechanics in chapter 8.)

What, then, are we left with in our experimental science? We have an extensive interlocking network of facts obtained by experimentation, theories serving to explain them but at the same time influencing their interpretation, and additional experimental data obtained in testing the validity of the theories. The details of the common practice followed in gathering the data and constructing explanations are often referred to as "the scientific method." If this phrase encompasses no more than taking extraordinary care in the assembling of facts, being as objective as possible in their interpretation, constructing falsifiable theories to account for them, and going out of our way to submit the validity of these theories to the most severe tests by repeatable and publicly accessible experiments (not necessarily comprehensible to the general public), rather than following rigidly laid-out procedures, it encapsulates what we do and what Steven Weinberg calls the "art of science."[18] Perhaps Percy Bridgman said it best, defining the scientific method as "using your noodle and no holds barred." The only problem with this catholic phraseology is that it makes the scientific method out to be nothing but common sense. So it may

---

[18] Steven Weinberg, *Dreams of a Final Theory* (New York: Pantheon, 1992), p. 131.

appear to us, who are completely used to it; but the historical fact that it arose only in Western civilization, with a strong inheritance from the Greeks, should tell us that it is not simply the natural, intelligent way of gathering knowledge of nature. No other culture developed this way of thinking, and many people still reject such a method. Of course, this historical fact opens the door to those who would argue that scientific knowledge is culture-specific, that other cultures have their own way of knowing. This may be, but none of the other ways can compare to that of modern science in comprehensiveness, reliability, effectiveness, and coherence.

OUTLINE OF THE BOOK

This book consists of a number of more or less independent essays on various general topics in physics, whose common thread consists of the pervasiveness of probabilistic approaches and the central role played by mathematics, with the quantum theory of fields as the most basic description of reality.

My first chapter delves into the nature of theories. Analogies, theories, and models all serve different purposes in enhancing understanding, but theories are the primary tool, though they are not uniquely determined by the facts. Are there crucial experiments, and what is the role of intuition?

In the second chapter I examine what is meant by the *state* of a physical system. The state of a system of particles in classical mechanics is defined in terms of their positions and momenta—the Newtonian equations of motion require these data as initial conditions; only then will there be *determinism*, by which I mean that the present state of a system determines its state at all later (and earlier) times. Additional data, on the other hand, would, according to these equations, be redundant and could not be freely assigned. In spite of determinism, however, sensitivity to initial conditions renders most systems chaotic and, for practical purposes, predictable only in a probabilistic sense. The issue of microstates versus macrostates is introduced, together with coarse graining. Quantum mechanics is as deterministic as classical mechanics, but the state of a system is defined differently. If the quantum state is defined in terms of a "complete set of dynamical variables"—a concept that differs from its analogue in classical mechanics—the values (measurement outcomes) of other dynamical variables form nonsharp probability distributions. It is at this point that probabilities enter quantum mechanics. On the crucial issue of entanglement, quantum mechanics differs from other possible probabilistic theories, and this is where the quantum world becomes *weird*. While part of what is meant by entanglement would obtain for any probabilistic theory—simply the consequence of probability correlations, examples of which I describe—other parts go further and are caused by *phase correlations*. Since entanglement in that sense does not strike us as strange for waves, whereas their nonlocal nature is counterintuitive for localized particles, I conclude that much of the strangeness of the quantum world originates in the wave-particle duality.

Chapter 3 deals with the power of mathematics in physics. Though the stimulus came from Galileo, it was Isaac Newton who made mathematics the language and most essential tool of the physical sciences. For the last three hundred years, almost all our theories have been formulated by means of differential equations. What do we mean by solutions of these equations, and what do we do with them? Much more is involved than simply generating numbers to be compared with experimental data. Sometimes an oversimplified mathematical model can serve an important explanatory purpose; as an example, we discuss the Ising model of ferromagnetism. The theory of group representations is a particularly striking example of the power of an abstract mathematical concept that goes far beyond numerical calculations. What explains the "unreasonable effectiveness," in Wigner's phrase, of mathematics in the physical sciences? I argue that mathematics is not embedded in the structure of the universe but is instead our most efficient and incisive logical tool for understanding nature. An appendix contains a discussion of the Korteweg–deVries equation—an example of a nonlinear equation whose solution was stimulated by computers but which finally required mathematical analysis to account for the appearance of solitons.

23

Michael Faraday introduced the *field* concept, discussed in chapter 4, in order to avoid action at a distance, a program that was carried to its logical conclusion by Maxwell. Formulating the laws of electromagnetism in terms of partial differential equations, Maxwell let the field at each point be determined by the field at neighboring points. Since the 1920s, we have *quantum fields*, operator-valued instead of number-valued functions. Particles emerge from quantum fields as a result of Fourier analysis together with quantization of each elementary "oscillator." They are the result of a remarkable property of harmonic oscillators—their quantum-mechanical spectrum has equal spacing between all energy levels. Once interactions are introduced in the theory based on the Dirac and Maxwell equations, positrons, in addition to electrons, emerge naturally, and the masses and charges of the particles are changed by renormalization, a procedure which removes the infinities that unfortunately beset the theory but which would have to be performed in any event. Similar methods are followed in other quantum field theories, with similar results. We conclude that both particles and waves—the latter from the description of the low-energy behavior of these particles by the Schrödinger equation—emerge in a natural fashion from quantum field theory. An appendix contains some details of the quantization of harmonic oscillators.

Chapter 5 is devoted to symmetries, which, in one form or another, have always played a role in physics. Ever since the time of Newton, however, the focus of attention has changed from symmetries in the observable facts, such as the orbits of planets, to symmetries in the underlying *laws*, from symmetries in the solutions to symmetries in the equations. We examine the case of mirror symmetry in some detail, including the discovery of violations of parity conservation in the weak interactions. Form invariance of equations under active and passive rotations, as well as under Lorentz transformations, are discussed, together with the necessary definition of vectors and tensors. Noether's theorem is introduced, which shows that the most powerful conservation laws are consequences of invariance under translations and rotations. A brief introduction to group theory follows, taking up the important applications of group representations in quantum mechanics and in particle theory. The chapter ends with the use of local gauge invariance as a method of generating fields and their equations.

Chapter 6 is concerned with causality and probability. What is meant by causality, and how can we tell the cause from the effect? The crucial issue for this distinction lies in the fact that the cause is under our *control* (the meaning of which is examined), but the effect is not. As a matter of universal experience it is found that the effect comes after the cause, a time-ordering whose particular consequences, together with the special theory of relativity, are frequently used in physical arguments, both in classical and quantum physics. This is the case even though quantum mechanics lacks causality: not every event can be assigned a cause; instead there are probabilities. I critically describe the frequency theory of probability as well as Popper's attempt to avoid frequencies. A number of the counterintuitive properties of any probabilistic theory (such as statistical mechanics, for example) are discussed. Here we find some of the same "strangeness" that is often claimed to be a special feature of quantum mechanics. For example, the probability for a system to be in a macrostate G_3 at the time t_3 if it was in the state G_1 at the time t_1 depends upon whether the system was observed at some intermediate time t_2, even if the results of that observation are ignored. Similarly, there is no time-reversal invariance, even if the underlying "flow" is time-reversal invariant.

Some of the issues taken up in chapter 6 naturally lead to the problem of the *arrow of time,* the subject of chapter 7. The thermodynamic arrow of time—the second law of thermodynamics—is discussed and its origin analyzed. This, however, is not the only arrow of time relevant to physics. There are also the causal arrow, defined in the previous chapter by the time direction from cause to effect; the cosmological arrow, defined by the expansion of the universe; the cognitive arrow, defined by our psychological experience of the unidirectional flow of time; and, finally, there is the direction of the time parameter that forms the fourth dimension in Minkowski space. Examining the relation of these five arrows of time to one another, we find that, given the causal arrow

25

as pointing to the future by definition, the thermodynamic arrow will necessarily agree with it (i.e., will point causally forward and the entropy of isolated systems will *increase*); the direction of the cognitive arrow, I argue, being determined both by the causal arrow and the thermodynamic one, therefore also necessarily points forward. The role of the fifth arrow of time, the direction of the time axis in four-dimensional space-time, is to assure, via relativity, that the causal arrow—and thus also the first two—are universal in space and time. The cosmological arrow, which is sometimes claimed to be tied to the thermodynamic one, does not appear to me to be connected to any of the others, which implies that we could, without contradiction, experience the universe to be either expanding or contracting.

The final chapter deals with the "weirdness" of quantum mechanics, beginning with the problems posed by the *Gedanken* experiment of Einstein, Podolsky, and Rosen, as well as Bell's inequality and Schrödinger's infamous pitiable cat. Questions about reality, too philosophical for most physicists' comfort, are here impossible to avoid. My view is that reality at the everyday level has to be distinguished from reality at the submicroscopic level. The most suitable description of the latter is given in terms of the quantum field. If our intuition, formed by macroscopic experiences, has difficulties with such a formulation, very well, then. We physicists should not flinch from accepting the world as we find it.

A brief epilogue picks up the thread of the book.

Theories

THE DIFFERENCE between a tentatively offered "explanation," by what may amount to no more than an ad hoc assumption, and a real explanation is the construction of a theory. Ad hoc hypotheses often serve a valuable temporary function, but in the long run they cannot take the place of a more structured explanation. When T. D. Lee and C. N. Yang solved the tau-theta puzzle by assuming a breakdown of parity conservation in the decays of these particles, skillfully avoiding ad-hockery by showing that the same effect occurs in other weak interactions (see chapter 5), their seminal solution did not become an explanation until that nonconservation was incorporated into a new version of the equations of the theory of beta decay and those of the other weak interactions. A few years earlier, the apparent contradiction between the long lifetimes of the newly discovered "strange particles" like the Λ and the Ξ on one hand, and their "copious production" in high-energy collisions on the other—which is why they were considered strange—had at first been "explained" by the fact that they were always produced in pairs, never alone (an observation referred to as *associated production*), but this did not rise to the level of a real explanation until the "strangeness quantum number" was introduced and hypercharge became an integral part of new field equations. More recently, the need to understand why the tau has never been observed to decay into a muon or an electron, or the muon into an electron, emitting a photon at the same time, led to the introduction of conserved lepton *flavors*. But although these flavors have now been incorporated into the standard model of elementary particles, there are still practicing high-energy physicists who regard flavors as merely a codification rather than a real explanation.[1]

The purpose of a theory is to lead to an understanding of the phenomena it deals with by connecting them, preferably via a

---

[1] See M. L. Perl, "The Leptons after 100 years," *Physics Today,* October 1997, p. 34.

causal chain, to other phenomena already understood, but it will serve that purpose only for those who are able to internalize it and make it, at least to some extent, part of their own intuition—that is, for those who understand the theory. A person may comprehend all the mathematical symbols of the theory of relativity, for example, and yet utterly fail to understand it. The process of thorough internalization and intuitive acceptance transforms a large theory of wide generality into a *paradigm.* This is exactly the sense in which some physicists—including very prominent ones like Richard Feynman—have alleged not to *understand* quantum mechanics, even though they have been using it continuously, perhaps making important contributions along the way.

There are, of course, many different kinds of theories with various degrees of generality and applicability. Apart from Einstein's distinction between those of principle and those of constructive mechanisms, it is useful to distinguish between *general* theories, such as the special and general theories of relativity, Newton's mechanics, or quantum mechanics, on one hand, and *local* ones, on the other.

Most of the theories produced by working physicists are of the local kind, designed within the framework of a more general theory but dealing with a restricted class of phenomena. Such are the theory of nuclear reactions, the BCS theory of superconductivity, the theory of beta decay, and the theory of stellar evolution. Whole subareas of physics are local theories with their own local paradigms: fluid dynamics, acoustics, solid-state physics, QED. All operate within the frame of a more general theory, sometimes based on approximations applicable only within their specific domain. Acoustics, for example, is a part of the theories of fluids and thermodynamics, with special approximations based on the assumption of small pressure variations; QED is a quantum field theory restricted to electrons in interaction with the electromagnetic field. Sometimes theories arise autonomously and are later subsumed by others, thus becoming local parts of a larger theory. Thermodynamics is one such example.

When the modern theory of heat arose in the early nineteenth century, it began as an independent field with its own laws. Only

28

when James Clerk Maxwell, Josiah Willard Gibbs, and Ludwig Boltzmann connected it to Newtonian mechanics by means of their new statistical approach were the laws of thermodynamics seen as consequences of Newton's laws governing the behavior of molecules. The resulting transformation of the second law into a derived probabilistic statement rather than a fundamental one like the first law, though of little experimental consequence, was so intellectually jarring that even Max Planck had a very hard time getting used to it.

Other distinctions between types of theories are also important for our understanding. Consider Newton's explanation of Kepler's laws, which consisted of two parts: the laws of motion, and the universal law of gravitation. The first part forms a large framework for the description of motions of all objects under the influence of forces of any kind, applicable to the trajectories of rockets in space as well as to the movements of molecules in a gas. The second part consists of a law describing the action of a specific force, gravity; revolutionary as this law was in its generality—it makes both the apple fall and the Moon orbit the Earth—its applicability is much more specific than the laws of motion. Newton's division of the question "how does object A move under the influence of object B?" into two parts—"what force does object B exert on object A?" and "what influence does this force have on the motion of A?"—has been enormously influential and has remained a paradigm in physics ever since. This division of the problem of how objects influence one another's behavior into a general equation of motion under arbitrary forces and, separately, a law governing the specific force involved, vastly extended the scope of his laws—we would now find it strange indeed to have different laws of motion depending on the specific forces involved—and it has persisted through later modifications, such as the special theory of relativity, blurred only by Einstein in the general theory of relativity. There, motion under gravity became simply a consequence of geometry and the equations of motion are not separated from force laws. The same dichotomy was also retained in quantum mechanics, the only difference being that the equation of motion is replaced by the Schrödinger equation and the force in each spe-

29

cific case is determined by a potential whose physical origin has to be sought elsewhere.

THEORIES ARE NOT UNIQUE

Though anchored firmly by empirical facts, theories are not uniquely determined by the experimental data; they are the product of physicists' imaginations. It follows that it ought, in principle, to be possible for different theories to account for the same set of observations. If we should ever be able to contact an alien civilization on a planet of Alpha Centauri, the chances are that their physical theories would not coincide with ours, though they would, of course, not contradict them, either. We do not have any examples of two nonequivalent theories covering the same ground,[2] except perhaps the standard formulation of quantum mechanics and David Bohm's version with hidden variables. We can explain this simply by recognizing that once a theory gets a good foothold, we are unlikely to be interested in another one that covers the same data with the same predictions and which is, therefore, experimentally indistinguishable from the first.

Can we really imagine the existence of a large overarching theory, equally valid as and competing with the Newtonian laws? I think we could, if we remember that no such theory accounts for the data *exactly*. In order to predict precisely the motion of a baseball by means of Newton's laws, we have to make all sorts of corrections for friction, air pressure changes, variations of gravity, etc.; other laws of motion would no doubt require different kinds of corrections. (We could use general relativity and make the needed approximations starting from that point.) Inconvenient as it certainly would be, we could even begin from the equations of fluid dynamics for a medium of very low Reynolds number and make the required large corrections to what would in effect

---

[2] The versions of quantum mechanics by Heisenberg and Schrödinger do not count; they were quickly shown to be equivalent.

be the Aristotelian laws of motion—if the aliens on Alpha Centauri lived in a highly viscous medium, they might well develop physical laws of that kind. I would therefore caution any scientist against assuming that our laws and theories are intrinsic parts of nature, rather than our way of describing and understanding what is out there.

How Experiments Decide between Theories

If theories are the physicist's primary explanatory tool and the facts found by experimenters are to some extent contaminated by theoretical interpretation, how can we assert that experimental results are the foundation stones on which the theories rest? Can a theory be *proved* by experiments or observations?

Clearly, in order for a proposition or an equation to be a candidate for a meaningful scientific theory, it must, in principle, be verifiable. There was a time when interpreters of science believed that the probability of a given theory's correctness could be ascertained by a well-defined "inductive logic": the more instances of agreement between the implications of a theory and experimental results, the more probable that it was right. Although there is, of course, an element of truth in this idea, no one has succeeded in making any sense out of the supposed probability involved here or in assigning it a numerical value. The most influential counterproposal was Karl Popper's: what matters most is whether a theory is *falsifiable*, not whether it is verifiable. But this idea has its limitations as well.

We have to distinguish between two kinds of criteria: tests to decide whether a theory is scientifically meaningful, and tests to determine its acceptability. For the first purpose, Popper's criterion of falsifiability is enormously valuable. If there is no way, even in principle, to devise a test that a proposed theory could fail and be discarded, it has no scientific meaning. Methods for verifying a theory are usually easy to come by; the theory, after all, was invented to account for certain observed phenomena. But, as in the case of psychoanalysis, it may be impossible to find a trial it could conceivably fail; there is always a possible reinterpretation that will save it. Nevertheless, even the falsifiability criterion is not foolproof. The facts used in an alleged falsification usually are contaminated by interpretations that depend on some theory; the falsification is therefore to some extent context dependent. Furthermore, the theory may be salvageable by minor ad hoc modifications, a deplorable but not unheard-of practice. Even Einstein once succumbed to that temptation, introducing the cosmological constant to prevent general relativity from predicting an expansion of the universe, for which there was no evidence at the time; after the discovery of the expansion by Wirtz and Hubble, he looked back upon this episode as the worst blunder of his life.[3]

All such cavils notwithstanding, Popper's central idea is still the best criterion I know for deciding whether a theory put forward has scientific meaning; in particular, it has to allow for clear-cut tests to decide between it and rival proposals. On the other hand, as a criterion for whether a given theory is right, falsifiability is patently useless. Certainly, every new theory must be subjected to the most rigorous attempts to prove it wrong, not just to find it right, and experimenters are to be strongly encouraged to try their best to search for discrepancies with theories rather than simply seeking agreement; but theories are primarily invented to be productive of new ideas and further understanding, not simply to avoid being false. If a new theory leads to important predictions, these predictions have to be tested, and their verifications are rightly regarded as triumphs. When Julian Schwinger calculated the anomalous magnetic moment of the electron by means of his new renormalization procedure for quantum electrodynamics and his result agreed with the previously measured value to better than a few parts in 10^5, he had found a dramatic confirmation of the theory, and QED became the hottest thing in town. Similarly with the discovery of the Ω^-, which greatly bolstered everyone's confidence in Murray Gell-Mann's *eightfold way*. But there

[3] Recent astronomical observations, however, seem to indicate an acceleration in the expansion rate of the universe, which would require the reintroduction of a cosmological constant, though not necessarily of the same size Einstein contemplated. Perhaps it wasn't such a blunder, after all.

is no convincing method of assigning a numerical *probability* of the correctness of a theory on the basis of such verifications. Of course, every successful outcome of a test at the same time constitutes an instance in which the theory *might have failed*, though not necessarily fatally. For example, if no method of calculation is known to predict the exact mass of a particle implied by a fundamental field theory, failure to find it need not be held against it, at least temporarily—witness the Higgs boson.

When it comes to the importance of confirming evidence for a theory, I think it is fair to say that physicists generally put a higher premium on *predictions* of as yet undiscovered data made by a new theory than on *postdictions* of experimental data already known.[4] There is no intrinsic reason to be more impressed if a theory predicts a new effect subsequently discovered in the laboratory than if it postdicts, or explains, one already known but hitherto not understood, unless this explanation is ad hoc. The justification for this preference—to the extent that it exists—may be that many physicists expect theorists to be clever enough to tailor their proposals to accommodate known facts; only when they go out on a limb in making a prediction do they run a serious risk of failure. In the case of Schwinger's calculation of the anomalous magnetic moment of the electron, it was, of course, clear that he could not have specifically devised his procedure to pass this test—it was surely not ad hoc—so the agreement impressed everyone mightily, even though it was a postdiction.

ARE THERE CRUCIAL EXPERIMENTS?

Popular accounts of science often dramatize stories by depicting some crucial experiment which confirms a theory once and for all and destroys a rival theory in the process. On the other hand,

---

[4]However, Steven Weinberg argues persuasively that the successful postdiction of a shift of the perihelion of Mercury was a more convincing piece of evidence in favor of Einstein's general theory of relativity than the sensational prediction of the bending of starlight by the sun; see Weinberg, *Dreams of a Final Theory* (New York: Pantheon, 1992), p. 96.

there are commentators like Pierre Duhem who deny the existence of such experiments, claiming they are all subject to assumptions and interpretations, which may change in the light of other data and theories. "[T]he physicist is never sure he has exhausted all the imaginable assumptions. The truth of a physical theory is not decided by heads or tails."[5]

A telling example given by Duhem is the experiment by Léon Foucault, demonstrating that the speed of light in water is lower than in vacuum, which at the time was interpreted as a crucial test in favor of the wave theory of light as opposed to a corpuscular model. Duhem argues that while it is easier to account for Foucault's result by means of Fresnel's theory than by Newton's, there was no reason why it might not be possible to devise a more sophisticated particle theory of light that could accommodate itself to the experimental data on the speed of propagation in media. As Louis de Broglie points out in his preface to a later edition of Duhem's book, in the very year, 1905, when it was written, Einstein published his paper on the photoelectric effect, proposing just such a theory—light as consisting of photons—which is perfectly compatible with Foucault's experimental result, proving Duhem to have been right.

Augustin Fresnel's diffraction experiment, however, constitutes a case with the opposite conclusion. To his own astonishment, he found that the shadow of a small circular disk contained a bright spot in its center, precisely the "absurd result" his opponent Siméon-Denis Poisson had calculated to be implied by the wave theory. As the somewhat oversimplified story goes, Fresnel's discovery convinced even Poisson, after which no viable theory of light could deny its wave nature.

I have juxtaposed the Foucault and Fresnel experiments to demonstrate that a well-designed experiment can at best decide between two very specifically stated theories, not between two general paradigms. While it is certainly an exaggeration to say that Foucault's experiment was a crucial test serving to decide be-

[5] Pierre Duhem, *The Aim and Structure of Physical Theory* (Princeton: Princeton University Press, 1991), p. 190.

tween wave and particle views of light, it did disprove Newton's specific theory. That a particle theory could arise again in spite of the observations of Fresnel's bright spot and Foucault's slowing of light in water is characteristic of the quantum theory, which combined waves and particles. Duhem's criticism of crucial experiments is therefore well taken up to a point. We should always be aware of the fact that test results are subject to interpretation in the light of other theories, that they might be circumvented by modifying the theory on trial.[6]

CONVENTIONALISM

The ever-present temptation to *save the phenomena* by means of ad hoc devices is one aspect of the argument advanced by the school of *conventionalism*, of which Henri Poincaré was a prominent, though cautious, member; the adherents of this school contend that large parts of our theories are conventional in nature. This is not necessarily to say that they are arbitrary, but they may be chosen because they are convenient. If this idea is not pressed too far—but philosophers do have a habit of pressing things too far—it is difficult to deny that there is some truth to this point of view.

Consider the example of the heliocentric solar system. In what sense can we claim that Copernicus was *right* and Ptolemy *wrong*? We can certainly describe the system in a reference frame in which the earth is at rest in the center, but such a description would be extremely cumbersome and inconvenient. We know that the adoption of an inertial system to view the sun and the planets has enormous advantages; can the heliocentric choice then be regarded as "arbitrary"? The very fact that describing the solar system with the sun at rest at the center, and explaining the planetary orbits by means of the Newtonian laws of motion in such an inertial system,

---

[6]See Weinberg, *Dreams of a Final Theory*, pp. 90–107, for an instructive discussion of the observations that are often cited as crucial pieces of evidence in favor of the general theory of relativity and against Newton's theory of gravity.

is very much simpler and more convenient than to describe it in a coordinate frame in which the earth remains at rest, says important things about nature. As Poincaré observes, "It is true that it is convenient, it is true that it is so not only for me, but for all men; it is true that it will remain convenient for our descendants; it is true finally that this cannot be by chance."[7]

The Newtonian laws of motion themselves are a more specific example of certain conventional aspects of our theories. Not only does the concept of *force* play an important role in them, but it is sometimes claimed these laws *define* what is meant by a force. The special theory of relativity, where a *Newtonian force* is defined to be one that obeys Newton's second and third law, seems to confirm this. However, we also know that we are intuitively aware of what is meant by a force: we can feel it; furthermore, its magnitude can be measured not only by the acceleration it produces, but independently by its role in statics—stretching a spring, etc. So to call it a convention is at best partially right. That the concept of inertial mass may be regarded as conventional rests on somewhat stronger grounds: it is surely defined by Newton's second law. Yet even in this instance, what appears as a definition turns out to have a great deal of empirical content. The mass enters into the definition of momentum, and the law of conservation of momentum plays a central role in dynamics. That the inertial mass then turns out to be equal (or proportional) to the gravitational mass appears to be an accident, an "accident" that later became a foundation stone of the general theory of relativity, in which the mass disappears altogether, enabling motion under gravity to be reduced to geometry. We see once more that if the concepts of mass and force are defined by the Newtonian laws of motion, the fact that these definitions turn out to be as fruitful as they are says much about the structure of nature.

This persuades me that there is a certain amount of truth to the contention that our theories contain conventional elements. Still, because these elements are limited in scope and hard to disentan-

[7] Henri Poincaré, *The Foundations of Science* (Lancaster, Pa.: Science Press, 1946), p. 352. The context of Poincaré's remark, however, was somewhat different.

gle from the directly empirical ones, it is unproductive and misleading to overemphasize them. Indeed, one aspect of this line of thought—Karl Popper calls it the "conventionalist stratagem"—is positively dangerous, because its consequences are sterile: if you regard the construction of theories as largely a matter of convention, you may be more tempted, when faced with a discrepancy between a prediction and an experimental result, to change the theory in an ad hoc fashion in order to make it fit the data (although Einstein did just that when he introduced the cosmological constant, without being a conventionalist).

Here are two prominent instances in the history of physics in which the "conventionalist stratagem" was avoided with great success. The experimental results in nuclear beta decay presented a puzzle. When a nucleus decays, producing a daughter of nuclear charge either one unit higher or lower, the emitted electron or positron would be expected to have a sharp energy equal to the difference between the rest energies of the initial and final nuclei, exactly like the line spectra of photons emitted by excited atoms.[8] Instead, the spectra of the emitted β-rays were found to be quite broad: there was, in most instances, a substantial amount of energy missing, leading Bohr to suggest the drastic solution of giving up the law of energy conservation. However, Pauli saved the day by proposing that the missing energy was carried away by an unobserved neutral particle of very small mass. (The mass had to be very small compared to that of the electron, because the spectrum of the emitted β-rays extended, so far as one could tell, all the way up to the energy difference between the initial and final states.) This new particle, dubbed "neutrino"—"little neutron"—by Fermi, had to have spin 1/2, because the electron or positron emitted has spin 1/2 and the difference between the angular momenta of the initial and final nuclei always was an integral multiple of Planck's constant—another part of the puzzle—able to interact with other

---

[8] In neither case are these energies completely sharp, of course, because the inital states, being unstable, cannot have exactly sharp energies. These widths, however, are usually too small to be experimentally resolvable.

particles only extremely feebly, since no indication of its presence was detected.

No wonder some physicists looked upon this as just the kind of ad hoc assumption to be avoided at all cost. But if the neutrino really existed, there should, in principle, be experiments, though difficult, that could detect it. Indeed, Cowan and Reines found this particle in inverse beta decay almost twenty-five years later. Even before its "direct" detection, however, the neutrino had manifested itself in so many other, less direct ways that there were few physicists who still doubted its existence—another instance of the coherence of our view of the world.

A second and perhaps the most striking case of navigating around the shoals of ad-hockery was T. D. Lee and C. N. Yang's discovery of the nonconservation of parity, which we shall discuss in more detail in chapter 5. In a spectacular demonstration of how to avoid the trap of having to introduce an ad hoc hypothesis they proposed that the two decay modes, one into two pions and the other into three, in what appeared to be a strange "parity doublet" called τ and θ, were, in fact, an instance of a parity non-conserving weak decay of only one kind of particle; at the same time, they suggested that the issue of parity conservation in other weak interactions such as beta decay should be experimentally tested.

MODELS

Theories, however, are not the only means of explanations used by physicists; there are also what we usually refer to as *models*. These generally differ from legitimate theories in being well understood to fall far short of describing nature. Nuclear physics, for example, is full of such models—three of them have been particularly important: Bohr's liquid drop model of the nucleus, and at a later date, the shell model and the collective model.

In the liquid drop model, Bohr disregarded the detailed constitution of the nucleus and was able to predict some of its important properties—especially those pertaining to fission—by considering it as analogous to a drop of liquid held together by its surface tension (in effect, an early version of what later became known as a collective model). Maria Goeppert Mayer, on the other hand, went to the other extreme and viewed the nucleus as consisting of a system of individual nucleons analogous to the electrons in an atom, with the important difference that the nucleus lacks a heavy center around which, and in whose central field, the lighter particles revolve. The nucleons, instead, are assumed to move in an effective force field that is the average of the forces exerted by all of them on one another. Since, according to quantum mechanics, the individual particles in such a mean field would then be arranged in shells at various energy levels, Mayer could explain many detailed nuclear properties, particularly spectroscopic data, on the basis of excitations of individual nucleons. Later, the collective model of the nucleus was introduced, which explains other nuclear properties on the basis of the aggregate rather than the individual behavior of the nuclear constituents, elucidating a number of data accounted for neither by the shell model nor the liquid drop. There is even a model that stands somewhere between the assumption of individual nucleons and that of the collective, in which certain nuclei are taken to be made up of a number of α-particles—an assumption justified by the particularly tight binding and stability of these four-particle clusters.[9]

That none of these theories serves to explain all the known properties of nuclei, that none of them was ever regarded as a full-fledged theory to be taken seriously in the sense of claiming "this is what a nucleus is like" to the exclusion of all other theories, transforms them into models. They complement one another, to this day standing more or less side-by-side, invoked in turn for the explanation of some data but not for others. Ultimately, they have to be justified in the context of quantum field theory of elementary particles, but as models they need not be rigorously derivable as local theories.

[9] For a variety of nuclear models, see, for example, J. M. Blatt and V. F. Weisskopf, *Theoretical Nuclear Physics* (New York: John Wiley & Sons, 1952); J. M. Eisenberg and W. Greiner, *Nuclear Models* (Amsterdam: North-Holland, 1970).

39

In the area of elementary-particle physics, phenomenological models are frequently employed for the purpose of explaining specific experimental results in simpler and more manageable ways than invoking a fundamental theory that might require more difficult calculations. The most general and prominent so-called model in the context of elementary particles is, of course, what at present is thought to be the most basic theory underlying all of particle physics—and yet it is called the standard *model*. Ignoring gravity, this fusion of the electroweak and strong particle interaction theories is highly successful in predicting a large number of the properties of elementary particles, but it is viewed as a temporary stepping-stone to a more general, all-encompassing "grand unified theory"—this is why it has the uncharacteristically modest name of a "model."

Models play important roles, as well, in astrophysics and the physics of condensed matter. The evolution of stars is explained largely by stellar models, including the "standard solar model," that are unrealistic in many respects but serve very well in making us understand certain specific aspects of the behavior of stellar interiors, such as a star's energy sources, its beginning, its evolution, and its ultimate fate. In solid-state physics, the Ising model of ferromagnetism is an example we shall discuss in chapter 3.

INTUITION

If conventionalists take one extreme view of the nature of physical theories, at the other extreme are those who put the primary emphasis on the *physical intuition* of their creators. The greatest physicists are said to be led to their important discoveries by using deep insight to sniff out the right kinds of questions to ask nature or to make a stab at the right answer. There can be no doubt that a refined intuitive sense can be of enormous value to a scientist. But again, let's look at what is meant by intuition.

How does an investigator decide to perform a particular kind of exploratory experiment? When confronted with new experimental facts, how does a theorist decide exactly where to look for an explanation among the body of accepted theories; how does he devise a new one, if necessary? Since there is no well-defined procedure of "inductive logic" that leads from the data to theories, to account for the genesis of a new theory becomes puzzling. It is here that physical intuition grows in importance, even though we cannot claim that all our significant theories were generated by such intuition. (Maxwell devised his theory of electromagnetism on the basis of a mechanical model that no one takes seriously any longer, and Dirac was guided primarily by esthetic considerations.)

I do not believe that what we call physical intuition forms a direct channel into the inner workings of nature. Such intuition seems to me, rather, to be based fundamentally on a very thorough internalization of all the concepts and implications of an existing body of knowledge and a feeling for exploration beyond its edges, coupled with a fine sense of what might be expected in a new region if the existing paradigm holds. This is where Thomas Kuhn's notion of the paradigm plays an important role; beyond the simple statement of a theory, it encompasses the entire penumbra of its implications and its approach to problems and questions. The more thoroughly it becomes a physicist's way of thinking, the more it becomes part of his intuition.[10] Today's physical insight has to include, when necessary, the entire quantum mechanical approach, but we appreciate Faraday's intuition even though he worked before the discovery of the quantum theory, and we credit Rutherford with great intuitive sagacity in his discovery of the nucleus, even though it was ultimately based on what we now know to be inapplicable physics. The successful experimenter, searching for a new discovery, uses her explicit and implicit intuitive knowledge of what is already known both in the design of the experiment and in the analysis of the results; taking data is the least of her jobs. And when a theorist proposes a successful new theory or a

---

[10] Anybody who has taught a course in introductory physics knows how hard it is to develop beginning students' intuition in Newtonian mechanics. We are all, it seems, natural-born Aristotelians. Physicists' intuition in classical mechanics has to be *learned*.

novel mathematical formulation to account for some set of unexplained data within a given paradigm, he is guided by his physical
intuition; only after the proposal is formulated can he and others
check its implications to see whether it is in accord with other facts
and whether its predictions are verified. The great leaps of imagination that lead to the development of entirely new paradigms—
Newton's laws of motion and his law of gravitation, Maxwell's
electromagnetism, Einstein's special and general theories of relativity, Schrödinger's and Heisenberg's quantum mechanics—are
much harder to pin down; "physical intuition" here surely is an
inappropriate nomenclature.

We are going to look now at some important specific issues
that arise in our theories, beginning with the definition of what
is meant by the *state* of a physical system.

The State of a Physical System

IF THE STATE *of an isolated physical system—such as the universe—is known at one time, its state is determined for all later times*: this statement is the essence of what is meant by determinism. (We might add it is the business of physicists both to find means of ascertaining what that state is, at any given time, and to discover the laws that allow us to determine the state at any later time.) But the question is, what do we mean by "the state of a physical system"?

THE STATE OF A CLASSICAL SYSTEM

In the context of classical mechanics, we all know that the state of a system of point particles—if the forces on them are given—is defined by their positions and velocities (or momenta). To list only their positions is insufficient, and to list their positions, velocities, and accelerations is redundant in the sense that they cannot all be assigned independently. This definition of "state" is surely not a priori obvious. Every physicist knows that to specify the state of the system is to list all the particles' positions and momenta because Newton's equations, or Hamilton's—coupled ordinary first-order differential equations in the time for the positions and momenta—require a knowledge of these data at one time in order to determine their values at any later time. Only if the state is defined in this way is classical mechanics deterministic. These equations also imply that we are not free to specify the accelerations. If we lived in an Aristotelian universe—suppose we were immersed in an infinite fluid of very low Reynolds number—in which the velocities are determined by the forces, the state of a system of particles would be given by their positions alone. We would still have determinism, but the velocities could not be freely assigned.

What is the state of an ideal fluid?[1] If its pressure and density are given, its state is defined by its velocity at every point, because Euler's equations are first-order differential equations in time for the velocity. Similarly, the state of an electromagnetic field without sources is defined by the values of the electric and magnetic fields everywhere because the Maxwell equations are of the first order, coupling both fields. For a classical system of moving charges, governed as it is by the Maxwell-Lorentz equations, the fields as well as the positions and momenta of the charges have to be specified.

To understand the limits of determinism even when the state has been appropriately defined, we have to be clear about how a system of particles needs to be set up in classical mechanics: all of them must have their initial positions and momenta precisely measured before the system is then let go. (Measurement and the preparation of a state are generally the same thing—a point that was emphasized particularly by John von Neumann in the context of quantum mechanics, but which holds in classical physics as well.) The state of the system can therefore be represented by a point in its phase space. If this were in fact practically possible, the future development would indeed be completely determined by the Newtonian (or Hamiltonian) equations of motion, and the state-point would describe a well-defined trajectory in phase space. In reality, of course, it is not possible to attain infinite precision, and we have to make do with small but non-zero errors, both in the initial positions and in the inital momenta. For the representation of the state in the system-phase space, this implies that at the start, the state is represented by a point about the location of which we can only say that it lies inside a small ball, whose diameter indicates the precision of the initial measurements. As time progresses, every point in the ball traces out a trajectory determined by Hamilton's equations—the Hamiltonian

[1] In thermodynamics, the state of a system such as a gas is generally defined by two of the three quantities—pressure, volume, and temperature—which are connected by the "equation of state." If two of them are given, the third is determined. But, its name notwithstanding, thermodynamics is not a dynamical theory: it predicts the state of a system at a future time on the basis of its present state only in rather general terms, such as "it will be closer to equilibrium."

44

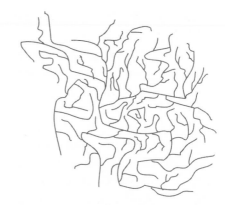

Figure 2.1. The Hamiltonian flow distorts the shape of the initial ball, though its volume remains unchanged.

flow—and the shape of the ball becomes increasingly distorted (fig. 2.1). Reassured by Liouville's theorem that the volume of the developing blob remains unchanged, we might expect that our particular system, which must remain in the moving blob in which it started, cannot get very far away from other points inside, and thus the error produced by the initial inaccuracies cannot get out of hand during the motion. This conclusion, however, is incorrect for most particle systems. As time passes, the initially nice compact ball will—Liouville notwithstanding—in almost all cases develop into a wildly distorted spidery shape with long thin tentacles, the distances between whose tips may be enormously large. This means two points that were initially very close together will generally end up far apart: two systems whose initial conditions were almost identical may wander off into quite different regions in phase space. To be more concrete: initially small inaccuracies become magnified, with the eventual result that the errors surrounding the future positions and momenta become so large as to render the attempted prediction useless: the system is *sensitive to initial conditions*, i.e., it is subject to the "butterfly effect"—the batting of a wing of a butterfly on the Amazon may alter the course of a tornado in Wisconsin—and there is *chaos*. This is the case even for such relatively simple systems as a single particle enclosed in an unsymmetric box, or a double pendulum, and certainly for all many-particle systems.

My brief recapitulation of some familiar results of classical mechanics serves as a reminder that prediction is, practically speaking, for many systems impossible, even though the state of a system of particles has been so defined that, in principle, the equations imply determinism. In addition to containing initial measurement errors, calculations of the future behavior of such systems, even by means of the most powerful computers, are always numerical and therefore subject to rounding errors at every time-step; their predictions are consequently limited to relatively short times. Weather forecasting is perhaps the most prominent example that has defied the early hopes of computer enthusiasts who were unfamiliar with, or had forgotten, results of mechanics already known to Poincaré.[2]

Nevertheless, there are some important physical systems, such as an idealized solar system, that are, at least to a very good approximation, deterministic even for purposes of practical calculation.[3] Furthermore, the issue of determinism remains an important one of principle. The fact that the definition of the *state* of a physical system varies with the equations of motion of the system makes the answer to the question "is nature deterministic?" depend on a definition, but it does not rob it of physical content. Nature could be so constituted that no definition of state would render it deterministic. However, in the context of quantum mechanics, as we shall discuss later, this realization does give the question a somewhat different cast.

---

[2] The subject of "deterministic chaos" arises for most systems subject to nonlinear evolution equations and was treated in detail by Poincaré about a century ago, though he did not use those terms. It has become fashionable in recent years, primarily because large-scale computers have made nonlinear equations, previously difficult to analyze, amenable to numerical investigation.

[3] It is good to remember, though, that even for the solar system the issue of long-term stability is not trivial. Laplace's worry about the issue led him to invent perturbation theory in order to check it out, but now the question can be answered by explicit computation, at least for about a billion years in advance. It turns out that the orientation of the Earth's axis is stable only because of the presence of the Moon. See J. Laskar, "Large-scale chaos and marginal stability in the solar system," in D. Iagolnitzer, ed., *Eleventh International Congress of Mathematical Physics* (Boston: International Press, 1995), pp. 75–120.

The original field in which physics explicitly abandons strict determinism and instead deals with probabilities is, of course, statistical mechanics. Since the complete microscopic state of a system of molecules making up a gas is never knowable, nor indeed of much interest, attention is focused instead on the macroscopic state of the system. To accomplish this shift of focus, we introduce *coarse graining*: the phase space is subdivided into cells whose size and shape are determined by the precision of our observations and the closeness of our attention in various regions of phase space. Most important, since it is of no interest which particular molecule of a fluid is located in a specified position, a coarse grain will contain all the microscopic states that differ from one another simply by exchange of identical molecules.[4] A *macrostate* thus defined corresponds to a large number of *microstates*; it is represented by a phase-space region of finite volume rather than by a point. Replacing microstates by macrostates in initial conditions will then result in the abandonment of strict determinism in favor of probabilistic predictions, based on a counting of the number of microstates contained in a given macrostate. This is the essence of classical statistical mechanics. Meanwhile, let us look at quantum mechanics.

QUANTUM MECHANICS

In quantum mechanics the state of a physical system is specified by its *state vector*, or more generally, by its *density operator,* and its equation of motion is the Schrödinger equation. The appropriate space in which to represent the state now is a Hilbert space rather than phase space. Since the Schrödinger equation is again a first-order differential equation, the future state of the system is fully determined if it is given at an initial time.[5] Quantum mechanics is, therefore, no less deterministic than classical mechanics. However, if the (micro)state of a classical system is given, the values

---

[4] We shall return to the subject of coarse graining in more detail.

[5] In the Heisenberg picture, the state is of course independent of the time, but the dynamical variables, as operators, obey first-order equations and are thus fully determined in the future if given at the present.

47

of all its dynamical variables are precisely specified. Therefore, if the values of a complete set of dynamical variables of a classical system are given precisely at one time, all its dynamical variables are completely determined for that time and for the future—the state itself, in fact, may be regarded as observable—whereas this is not the case in quantum mechanics: the state of a system does not determine all its dynamical variables precisely, nor is it an observable.

First of all, the phrase "a complete set of dynamical variables" of a system has different meanings in classical and in quantum mechanics. In the first instance it denotes a set of quantities, such as the particles' coordinates and momenta, whose values at one time completely determine the state of the system and hence the values of all of its other dynamical variables. In quantum mechanics, that phrase denotes a set of commuting operators whose spectra can be used as unambiguous labels on a basis (either literally or in the extended sense of generalized Fourier integrals) of the Hilbert space of the system, which implies that their simultaneous eigenvalues (or quasi-eigenvalues) may be used as unique (up to a constant factor) identification of a state vector. In a state labeled in this way, the values of other dynamical variables are determined, if they are determined at all, only as probability distributions. *It is at this point, and not through a lack of determinism, that probabilities enter quantum mechanics.* Classically, the position of a particle may *ideally* be precisely determined at the time $t = 0$; *practically*, it cannot be precisely pinned down. But (for a free particle) the position error $(\Delta x)_t$ at a later time t, which grows because its momentum cannot practically be specified precisely either, can be reduced to a value as close to the initial $(\Delta x)_0$ as the ingenuity of the experimenter and the quality of the equipment allow; quantum mechanically, on the other hand, the spread of the error has an irreducible minimum determined by the fact that, given the initial position accuracy, Heisenberg's indeterminacy principle sets a lower limit on the initial momentum error. The future position of a particle is therefore more uncertain than it was initially—the wave packet, though fully determined, spreads—and the particle, in contrast to the point in Hilbert space representing its state, cannot be described as having a well-defined trajectory; this is so because of the nature of the quantum state of the particle at one time, and not because quantum mechanics lacks determinism. Furthermore, the growing uncertainty of a particle's position takes place even for the simplest system imaginable—a single free particle—and not only for relatively more complicated systems as in classical mechanics.

Similar arguments hold for unstable configurations, the prototype of quantum-mechanical unpredictability. Consider a system consisting of two particles A and B, both interacting with a force center but not with each another. Suppose there is a state Ψ_1 in which particle A is in an excited bound state of energy E_1^A and particle B is in its lowest bound state of energy E_0^B. Assume the total energy $E = E_1^A + E_0^B$ is so high that if particle A were in its ground state with energy E_0^A, there would be enough energy to lift particle B into the continuous spectrum, i.e., to make it "free." Let Ψ_2 be such a state with the same energy $E = E_0^A + E_{free}^B$. If the system is initially in the state Ψ_1, it will remain there. However, in the presence of an interaction between particles A and B, the initial state Ψ_1 will evolve into a superposition that includes both Ψ_1 and Ψ_2. This means that at a later time, the state of the system will consist not only of both particles bound, as initially, but will contain a component in which particle B is free, i.e., in which the system has decayed. Even though the time development of the state is deterministic, the later state is such that there is an increasing *probability*—rather than a definite prediction — of finding the system disintegrated at the time t.[6]

The entry point for probabilities in quantum mechanics is therefore not nondeterministic predictions, but simply that dynamical variables are represented by operators, and noncommuting operators do not have simultaneous eigenvalues. If the value of one variable is fixed—by having been measured—the values of others have nonsharp probability distributions. That the result of this is an indeterminacy in predictions of the future behavior of the dy-

[6] This, for atoms and molecules, is the Auger effect; for molecules, it is called predissociation; in nuclear physics, internal conversion is an effect of a similar nature.

namical variables of systems is a secondary effect, one which quantum systems share with chaotic classical systems. (The origins of the unpredictability in the two cases are, however, quite different.) My point here is not to try to explain away the acausality of quantum mechanics but to pin its source down more precisely than is usually done. *Probabilities are a central feature of quantum mechanics because of the very concept of the quantum state of a physical system.*

The quantum state of a system of particles, if specified in terms of dynamical variables as precisely as quantum mechanics allows, i.e., represented by a state vector, is much more analogous to a classical macrostate than to a microstate. If we adopt coarse graining in a classical phase space based on a given choice of canonical variables, specifying a macrostate also determines other dynamical variables only in the form of probability distributions, as is the case for quantum states. In terms of dynamical variables, one should therefore not look at wave functions as analogous to microstates and density operators as analogous to macrostates. *There is, in those terms, no quantum analogue of microstates.* The futile search for hidden variables is precisely the search for such microstates. On the other hand, if we focus on the state of a system itself, rather than on its dynamical variables, the appropriate analogue of a microstate is indeed the state vector and that of a macrostate, the density operator.

To be more precise about the quantum-mechanical definition of a state, we have to make a distinction between point spectra and continuous spectra. The point eigenvalues of an operator[7] determine the corresponding eigenvectors uniquely (to within a constant factor). If a state is prepared by the measurement—even though it cannot be completely precise—of a dynamical variable in a neighborhood of an isolated point eigenvalue, the result can generally be assumed to be a unique state. Given this state, the value (measurement outcome) of the defining variable is, in principle, sharp, but those of all other dynamical variables that do not commute with the defining one form nonsharp probability distri-

---

[7] In the absence of degeneracy, or of a complete set of commuting operators if there is degeneracy.

butions. On the other hand, if a variable is measured in the continuous spectrum—such as the position of a particle—its values can *practically* be determined only within a certain error limit Δ, and the result is not a unique state. In that case, the state vector is unique (to within a constant factor) only if the (imprecise) values of other, noncommuting variables—such as the momentum—are also given. For such states, *all* dynamical variables have generally nonsharp probability distributions. If the measurement of the defining variable is idealized to give a sharp result, others are not determined at all. (In a quasi-eigenstate of a particle's position, its momentum is completely indeterminate.)

Let us take, as a simple illustration, a free particle whose state is represented by the configuration-space wave function

$$\psi(x) = \text{const.} \times e^{-x^2/(2\Delta^2)}.$$

Its position has the probability density distribution $\exp(-x^2/\Delta^2)$, which makes its location at the origin certain to within Δ. This particle is essentially at rest, but its momentum is uncertain by $\hbar/\Delta$, with a probability density distribution proportional to $\exp(-p^2\Delta^2/\hbar^2)$. However, the wave function

$$\psi'(x) = \text{const.} \times e^{-x^2/(2\Delta^2)-i k_0 \cdot x}$$

also represents a particle located at the origin within the error Δ and with the same probability distribution $\exp(-x^2/\Delta^2)$, but in this state its momentum is $\hbar k_0$, with an uncertainty $\hbar/\Delta$ and a probability distribution proportional to $\exp[-(p - p_0)^2\Delta/\hbar^2]$. So, a unique determination of the state requires the (approximate) measurement both of the position and the momentum.

We will, in a later chapter, discuss in more detail the concept of probability so crucial to the interpretation of quantum mechanics; suffice it to say at this point that the only interpretation that makes sense to me is that of *frequencies.* It can be applied, however, only when we can imagine a long (in principle, infinitely long) sequence of repetitions of the same experiment, or an ensemble of identical systems subject to similar conditions. An immediate consequence

51

of this interpretation is that the quantum-mechanical state vector never refers to, or completely describes, an individual physical system, but always refers to an ensemble (which is the reason why some physicists resist this interpretation). We should therefore say more accurately than before that in quantum mechanics, *the state of a system is defined as the ensemble to which the system belongs*, and this ensemble is represented by a vector in Hilbert space or by a density operator, depending upon the precision with which the ensemble is defined. The quantum state of a system is *not* a direct description of the individual system.

Entanglement

The issue of *entanglement* of distant particles, another characteristic of quantum states, opens up a separate set of questions. When specifying a classical state of several particles that do not interact with one another, their individual dynamical variables can be independently assigned, even if in the past they interacted. In quantum mechanics this is not necessarily so; there are the notorious phase correlations, which are difficult to grasp intuitively.

Recall Bohm's version of the EPR experiment.[8] Two spin 1/2 particles emerge as the decay products of a spin 0 parent, flying off in opposite directions. Since their total angular momentum must be zero, measurement of any spin projection of one of them allows us to infer that of the other, no matter how far away: they are entangled. If the vertical projection of the spin of particle 1 is measured and found to be *up*, a measurement of the vertical projection of the spin of particle 2 *must* yield *down*. But if it was a horizontal projection of the spin of particle 1 that was measured

---

[8] A. Einstein, B. Podolsky, and N. Rosen, "Can quantum mechanical description of physical reality be considered complete?" *Physical Review* **47** (1935), p. 777; N. Bohr, "Can quantum mechanical description of physical reality be considered complete?" *Physical Review* **48** (1935), p. 696. Bohm's version can be found in David Bohm, *Quantum Theory* (New York: Prentice Hall, 1951), pp. 614ff.

and found *right*, the horizontal projection of the spin of particle 2 must be *left*, and a measurement of its vertical component will, according to quantum mechanics, give the result *down* only with a probability of 50%.[9]

That this apparently strange situation is in part a direct consequence of the probabilistic nature of a quantum state will be discussed in more detail in chapter 6. But while probabilistic states necessarily lead to counterintuitive correlations, this does not exhaust the entanglement issue as it arises in quantum mechanics. If a several-component system can have a certain property in various ways made up out of the properties of the individual systems, probabilities become correlated. Suppose a red die and two white dice are thrown for a total of 7; given that the red one shows 1, the probability that one of the white ones shows 1 is $P_7(1, 1) = 2/5$; given that the red one shows 4, the probability for one of the white ones to show 1 is $P_7(1, 4) = 1$. (How do the white ones "know" what the red one showed?) If we pay no attention to the red die, on the other hand, the probability that at least one of the two white dice shows 1 (still given that the total is 7) is $9/15$. The showing of the red die and the appearance of a 1 among the white ones are correlated. Similarly, the states of the electron and the neutrino in nuclear β-decay are correlated, as they would be for any probabilistic decay theory. But there is no classical analogue of the *coherence* property of pure quantum states connected with the superposition principle. The state (ensemble) of the three dice showing a total of 7 by exhibiting 1 on the red die and 1 on one of the white dice cannot be decomposed into other states, one of which has 2 showing on the red die and 3 on one of the white dice, as the state of a circularly polarized light beam can be decomposed into vertically and horizontally polarized states. For this we need the coherence properties of waves.

Take a circularly polarized light beam, split in two, one half passing through a polarizing filter leaving it plane polarized hori-

---

[9] We shall return to a discussion of the EPR experiment in chapter 8. The discussion at this point is not meant to exhaust the implications and the significance of the subject.

zontally, and the other polarized vertically. If these two plane polarized beams are eventually brought together again, the resulting
beam will in general be unpolarized. However, if sufficient care
has been exercised—this is not necessarily practical, but I am interested in the principle only—so that the phase relations between
the two are preserved, the combined beam will again be circularly
polarized; the two beams were *coherent*, and their coherence was
preserved. This result does not strike us as weird or counterintuitive. Our intuition balks at entanglements of this kind only for
particles, because they are spatially confined, local, and individual. The state of a classical system of particles, either probabilistic
or causal, has no room for the concept of coherence.

This example—and, similarly, the formal demonstration of entanglement in terms of wave functions—shows that ultimately it
is the wave-particle duality inherent in the quantum theory that
is responsible for these counterintuitive effects. Even when we describe the state of a system in terms of particles, remnants of the
attached wave aspects remain and make their presence known.
Entanglement is then nothing but the familiar coherence of the
corresponding waves.

The issue of coherence is, at the same time, the defining difference between describing a state by means of a vector or a density
operator. When a dynamical variable A is measured,[10] the state
of the system after the measurement is the projection of the original state on the eigenstate of A corresponding to the measurement
outcome. (This is von Neumann's projection postulate.) What this
means is that the ensemble consisting of identical systems subjected to the same measurement and with the same (nondegener-
ate) outcome, is in a pure state, described by a state vector. However, the ensemble consisting of the systems subjected to the same
measurement irrespective of the outcomes—in other words, all the
systems are kept, no matter what the result of the measurement—is

---

[10] More precisely, when a complete set of commuting variables is measured and
the result lies in a discrete point spectrum, or if the imprecise result lies in a
continuous spectrum and the conjugate variables are also measured.

always in a mixed state and must be described by a density operator. The subensembles corresponding to the various measurement
results are *incoherent*.

For example, the splitting of a coherent beam of spin 1/2 particles by means of a Stern-Gerlach apparatus into two separate
beams, of spin down and spin up, respectively, does not by itself
constitute a measurement—the two partial beams, still coherent,
may be recombined to form a beam like the original one; each still
in a pure state, they are properly described by a state vector or
wave function, and their superposition is again a state vector. A
measurement (of a spin projection of particles) is performed only
when we identify particles in one (or both) of the beams and count
them; the two beams must now be described by a density operator.
Treating the constituents as *particles* is what destroys the coherence.

Measurement of course does not represent the only way in
which mixed states are produced in nature. Consider an idealized situation in which an infinitely massive unstable system
decays into two stable fragments of zero spin—like the α-decay
of a heavy nucleus into the ground state of its spin-zero daughter. If we suppose the state of the initial system to have been
pure (this is where the idealization comes in), so will be the state
of the emitted fragment (the α-particle). Even though the emitted fragment cannot have a completely sharp energy (because
the initial system, being unstable, could not have a sharp energy)
it is in a pure state describable by a state vector. On the other
hand, in a three-particle decay in which one of the fragments is
unobserved—as in β-decay, in which a neutrino is emitted—the
state of the observed fragment is mixed; the coherence is lost
with the unobserved third particle. The emitted electron *by itself*
is in a mixed state, even though the state of the electron-neutrino
system may be pure: the electron and the neutrino are entangled, and the entanglement is broken by any measurement on
the electron alone. The same is also the case for the α-particles
in the first instance if the emitting nucleus is not infinitely massive; the system consisting of the recoiling nucleus and the emitted
α-particle may be pure—the α-particle and the residual nucleus

are entangled—but the α-particle itself is necessarily in a mixed state of various energies, without coherence.[11]

The presence of coherences in quantum states, associated with the superposition principle and waves, is what primarily distinguishes these states from classical, even probabilistic ones. The quantum state of two spin $1/2$ particles with total angular momentum zero can be described as a superposition of states in which the individual particles have spin up and down, respectively, and, equally well, as a superposition of states in which they have spins right and left. This is what makes the EPR *Gedanken* experiment so counterintuitive, and this is what EPR interpreted as necessarily implying that quantum mechanics must be an incomplete description of reality. There is no classical analogue of the distinction between ensembles corresponding to pure states and those of mixed states, nor is there a classical counterpart to the coherent mixing (superposition) of pure-state ensembles, leading again to pure states—the superposition principle—in contradistinction to incoherent mixing, which leads to mixed-state ensembles. These particular features of quantum ensembles arise from the wave-particle duality.

The principal reason for our special unease with the correlations implied by entanglements—correlations which go beyond those inherent in probabilistic states of the classical kind—is that we are focusing on *localized* events, as is natural when states are described in terms of particles. For waves, on the other hand, we are quite used to these kinds of correlations and consider them normal. It should be clear from this that any kind of hidden-variable particle theory, such as Bohm's for example, which attempts to duplicate all the results of quantum mechanics, must necessarily be nonlocal, a result that was in fact proved by John Bell. Without strongly nonlocal features the theory could not possibly lead to the entanglements produced in quantum mechanics by its wave-particle

---

[11] The most striking experimental difference between the two kinds of decay is that the observed particles in α-decay are (almost) monoenergetic, while those in the three-particle β-decay are not—that's why Pauli postulated the existence of the neutrino—but this is not the main issue for our discussion at this point.

duality. Nonlocality is a natural characteristic of waves and does not offend our intuition, while for particles it is extremely counterintuitive and would be equally so in any hidden-variable theory in which the hidden degrees of freedom are those of particles.

SUMMARY

To summarize our discussion of what is meant by the *state* of a physical system, the most important point is that the meaning of this concept is not intrinsic to nature but determined by our mode of its description. We cannot expect anyone unfamiliar with details of physics, especially beginning students, to understand what is meant when a physicist refers to the state of a system. Indeed, the meaning of the term has changed in the course of history. This is not to say, of course, that we impose a state upon nature; it only means that we choose the mode of its description. The link between the situation "as it is" out there and our account of it is the act of observation or measurement. In some instances the outcome of a measurement may require a discontinuous change in our description—perhaps in aspects of it that refer to distant locations—but this does not necessarily imply a discontinuous change in nature itself. We shall return to this point in later discussions of probabilistic theories and the notorious "collapse of the wave function" in quantum mechanics.

Let's turn, then, to a discussion of the role that mathematics plays in physical theories.

The Power of Mathematics

"THE LANGUAGE of nature is mathematics," Galileo thought, and ever since, the formulation of physical theories has been almost invariably mathematical. But surely, mathematics is far more than a language; indeed, it serves both as a structural tool for physics and as a powerful instrument of thought. The very architecture of our science is determined in part by mathematical ideas. In this chapter I want to examine the possible reasons for a power whose striking effectiveness Eugene Wigner considered to be "unreasonable."[1]

A BIT OF HISTORY

Though modern science began with Galileo, it was Isaac Newton who determined the course of physics on its path of mathematization. After all, he specifically invented the calculus for the purpose of applying it—the differential calculus for his equations of motion, and the integral calculus in part for its use in applying his universal law of gravitation to extended bodies. For well over two hundred years after Newton, physics and astronomy provided a fertile soil for the growth of mathematics, and many of the greatest mathematicians made important contributions to the development of physics; we need only remember Daniel Bernoulli, Leonhard Euler, Jean d'Alembert, Pierre Simon de Laplace, Adrien Marie Legendre, Karl Friedrich Gauss, Karl Gustav Jacobi, William Rowan Hamilton, and Henri Poincaré to recognize the symbiotic relationship between the two sciences.

As a result of this intimate connection, the terms "mathematical physics" and "theoretical physics" were, until the beginning

---

[1] Eugene Wigner, "The unreasonable effectiveness of mathematics in the natural sciences," *Communications in Pure and Applied Mathematics* **13**, no. 1 (1960).

of this century, used interchangeably; only during the last hundred years, when an ever-growing specialization became the norm, did the two become separated, in part because many mathematicians, calling themselves "pure," became increasingly preoccupied with abstraction and disdained applications, in part because some physicists began to distrust a heavy emphasis and reliance upon more and more recondite mathematics. At the same time, the areas of mathematics utilized by physics proliferated. From Newton to Maxwell, it was primarily analysis that was used, and specifically ordinary and partial differential equations; then Einstein formulated his general theory of relativity in the language of differential geometry, developed mostly during the preceding half century, and Werner Heisenberg's and P. A. M. Dirac's quantum mechanics made heavy use of abstract algebra and operator theory, neither of which was familiar to most physicists at the time. Quantum mechanics eventually relied on the theory of groups and functional analysis as well, and quantum field theory led to further applications of new mathematics, including abstract versions of probability theory, and, most recently, the theory of knots.

The increasing emphasis on newly developed or unfamiliar mathematics caused theoretical physics to split into two branches: many theorists now work without using the more advanced of these tools and formulate phenomenological models that sometimes turn into local theories, closer to experimental data, relying on intuitive approximation methods. Others, the "mathematical physicists," work in a more rigorous mathematical mode, keeping abreast of new mathematical results, relying less on their physical intuition than on strictly proving mathematical conclusions that others either take for granted or are unaware of.

THE ROLE OF MATHEMATICS

What does mathematics actually do for us in physics? In the first place, it provides us with a language that is both unambiguous and universal, understood by scientists all over the world with the appropriate education (at the cost of being opaque to the vast

59

majority of nonscientists). To be more specific, let us look at some details.

On the most elementary level, there is the fact that almost all theories are expressed in the form of differential equations. These are either ordinary differential equations, such as Newton's equations of motion, or partial differential equations, such as Maxwell's equations of electromagnetism, the equations of fluid dynamics, those of acoustics, and the Schrödinger equation. In order to discover what the detailed predictions of these theories are, and to confront them with experimental results, the equations, together with appropriate initial and boundary conditions, have to be "solved," that is, methods have to be developed to express the functions (of time and space) that satisfy them in such a way that they can be numerically calculated for any value of their argument. Many mathematicians of the nineteenth century spent large parts of their lives investigating the most suitable procedures for doing this and tabulating the values, primarily obtained by series expansions, of those functions that occur most frequently as solutions of the differential equations of physics: Bessel function, Hankel functions, Legendre functions, hypergeometric functions, and many more to be found in reference tomes listing their properties and numerical values. These "special functions" can then be used to express solutions of many of the equations encountered and to furnish precise numerical predictions for all needed quantities subject to experimental tests.

The differential equations of our theories, however, always contain arbitary functions, expressing potentials, current distributions, indices of refraction, boundaries, and other details of the input. Solution methods had to be developed for all of them, which is how many of the differential equations, together with their boundary conditions and initial conditions, came to be replaced by integral equations. Finally, in order to be able to calculate actual numbers, appropriate approximation methods had to be found that allow us to compute the needed numerical results with high precision.

The use of mathematics in physics, however, goes far beyond the need to generate numbers for comparison with experimental data. The many books listing the properties of "special function," as you well know, contain not just numerical tables. They also show the results of detailed investigations by mathematicians, listing asymptotic values of various kinds for these functions and their analytic properties. The solution methods for the more general equations, such as integral equations, similarly lend themselves to such investigations, from which we obtain the most interesting consequences of the theories. Think of the procedures that allow us to infer, from the Maxwell equations, the different kinds of radiation patterns emitted by moving current distributions, or from Schrödinger's equation the differential cross section for the scattering of particles by one another and the bound-state eigenvalues and their eigenfunctions. Even though, in all of these cases, we again eventually end up with numbers which can be compared with experimental results, it is clear that the use of mathematics here goes beyond what could be accomplished by simply feeding the differential equations of our theories into a computer.

The Function of Computers

I remember once, many years ago, having a conversation with a prominent chemist who expressed the opinion, with which I emphatically disagreed, that there really was no use for theories any longer; all we needed to do was feed the data into a powerful computer and out would come predictions that could be checked. If theories served only the purpose of making predictions of future phenomena on the basis of known facts, as some scientists seem to think, computations could easily take their place. But enormously useful as computers may be, their function is limited by the very purpose theories serve—as tools for our understanding of the world. By this I mean not only that computers will never supplant grand theories like relativity or quantum mechanics, but their role will remain circumscribed even in much more specific situations, where we employ theories of local scope.

Take the behavior of fluids, for example. We know that a fluid consists of large numbers of molecules, whose behavior in collisions at the relevant low energies is well understood by classical mechanics. The whole field of fluid dynamics is a successful local theory based on the Newtonian equations of motion together with certain simplifying assumptions and approximations. Nevertheless, we still do not understand such a specific phenomenon as turbulence. Even if we had computers so powerful as to allow us to solve the equations of motion of all the constituent molecules of the fluid, being able to predict its behavior in all circumstances would explain nothing. Whatever form the eventual explanation of this phenomenon will take, simply being able to predict by means of computation is not the same as understanding, suggestive and stimulating as such detailed predictions may be in generating new theories.

The discovery of solitons provides another instructive example. The Korteweg–de Vries equation describing the motion of waves in shallow water had been around since the end of the nineteenth century, but since it is nonlinear it was not very well understood, except that it had solutions which were "solitary waves," simple heaps with one maximum, whose propagation velocities depended on their amplitudes—waves such as the naval architect John Scott Russell had oberserved in the Union Canal in Edinburgh. When Martin Kruskal and collaborators put this *KdV equation* on a computer in the 1960s, they discovered a new phenomenon: two such solitary waves of different amplitudes coming in from one side, with the faster one behind, would eventually meet as the faster began to overtake the slower; they would interact and oscillate—this is a nonlinear equation, not subject to the superposition principle— and then separate again, now with the faster one in front but with the shapes and amplitudes of both unchanged. Such waves that retain their identities in collisions, much like particles, were dubbed *solitons.*

Fascinating as this new phenomenon was, it was not explained by the computer printouts but remained puzzling until the Kruskal group and others came up with a mathematical analysis that was as surprising as the existence of the solitons itself. In a prime example of how much farther such analysis can take us than simple computation,[2] they were able to connect the nonlinear KdV equation to the (linear) time-independent Schrödinger equation in one dimension, with a specific family of potentials depending on a parameter t (the time in the KdV equation, but having nothing to do with time in the Schrödinger equation): the shapes of the potentials and their variations with t precisely describe the motion of the solitons if we require that, as t varies, the energy spectrum of the time-independent Schrödinger equation remain unaltered and the reflection amplitude undergo a simple linear change of phase. The connection between the KdV and the Schrödinger equations—though this has nothing to do with quantum mechanics—can be exploited to solve the initial-value problem for the former by means of the latter: given the shape of the wave at the initial time, calculate the reflection amplitude and bound states of the Schrödinger equation with (the negative of) this wave as the potential, then solve the "inverse problem" (by known methods) to find the potential from the knowledge of the same spectrum and the changed reflection amplitude, and thus the wave at a later time. Thus the "inverse scattering method" for solving the initial-value problem of a nonlinear equation by linear procedures was born.

The moral of the story is that the computer played an important role as a stimulant, leading to the discovery of a hitherto unknown physical (and mathematical) phenomenon occurring in a variety of contexts. But only real theoretical analysis could provide an explanation.

The computer has become especially important in connection with nonlinear effects in physics. In contrast to most of the earlier equations of classical mechanics, those of classical electromagnetism and quantum mechanics are linear, which is why for about a century physicists and mathematicians have been preoccupied with linear equations. The advent of the computer has made it possible to deal numerically with nonlinear differential equations,

---

[2] For more details, see the appendix to this chapter. An extensive guide to the literature can be found in "Resource letter Sol-1: Solitons," by A. Degasperis, *American Journal of Physics* **66** (1998), p. 486.

and as a result, nonlinear phenomena have come to the forefront of our attention, have indeed become quite fashionable. Still, the computer can do no more than serve as a stimulus; no matter how powerful, it will never take the place of explanatory theories.

This is not to say, however, that computers cannot serve an important role in guiding our intuition and helping our understanding of complicated natural phenomena. My earlier claim that even if we could solve the equations of motion of the constituents of a turbulent fluid on a computer we still could not explain turbulence does not mean that such computations would not be, within limits, very useful to our understanding of the phenomenon. It could assuredly lead us to an intuitive feeling for the importance of certain parameters and for the effects of changing them, especially in graphic representations. As some mathematicians have already discovered, the computer can play a role analogous to that of an experimental apparatus, and for physicists, too, it can lead to a form of understanding of natural phenomena analogous to that which many experimentalists acquire directly from manipulating their equipment, without any help from theory. It is clear, nevertheless, that such understanding would be restricted in scope.

MATHEMATICAL MODELS

Since the introduction of quantum mechanics in the 1920s, physicists have expected to be able to understand ferromagnetism on the basis of the magnetic moment interaction of the spins of the atoms making up magnetic materials. But could quantum mechanics together with thermodynamics account for the existence of a phase transition? Even if it should turn out to be too difficult to calculate the Curie temperature for individual ferromagnetic materials with any precision, could one at least explain the existence of a temperature above which ferromagnetism abruptly disappears?

Suppose the atoms are regarded as fixed at infinitely many sites on a stationary regular crystal lattice and their magnetic moments are assumed, for simplicity, to interact with a uniform external magnetic field in the z-direction and with their nearest neighbors only. The Hamiltonian for this system can be written approximately as

$$\mathcal{H} = -\tfrac{1}{2}J{\sum_{i,j}}' \vec{\sigma}_i \cdot \vec{\sigma}_j - \mu H \sum_j \sigma_j^z,$$

where the $\vec{\sigma}_i$ are the spin operators for the atoms, J denotes their interaction strengths, μ their magnetic moments, H represents the externally applied magnetic field, the sums run over the entire (infinite) lattice, and the prime on the first sum indicates that only nearest neighbors are included (this is where the approximation comes in). The thermodynamic properties of the system can be calculated by means of the partition function

$$Z = \sum_n e^{-E_n/kT},$$

where the sum runs over all the eigenvalues E_n of $\mathcal{H}$, and one is primarily interested in the limit of $H \to 0$, i.e., the behavior when the magnetic field is turned off. The problem, however, of evaluating this partition function and analyzing its properties as a function of the temperature T is impossibly difficult because of the lack of commutativity of the components of $\vec{\sigma}_i$, particularly since no singular behavior of Z can be expected for finite lattices. In order to cut through this problem, Ernst Ising introduced the simple idea of deleting the x and y components of the spin operators, so that all that is left are the z components.[3] Since these commute, they can be simultaneously diagonalized and replaced by their eigenvalues $\pm\tfrac{1}{2}\hbar$.

The realistic problem of ferromagnetism has therefore been replaced by a mathematical model, in which the Hamiltonian is given by[4]

$$\mathcal{H}_{\text{Ising}} = -\tfrac{1}{8}J{\sum_{i,j}}' \sigma_i \sigma_j - \tfrac{1}{2}\mu H \sum_j \sigma_j,$$

---

[3] Ernst Ising, "Beitrag zur Theorie des Ferromagnetismus," *Zeitschrift für Physik* **31**, p. 253, 1925. See also G. F. Newell and E. W. Montroll, "On the theory of the Ising model and ferromagnetism," *Reviews of Modern Physics* **25** (1953), p. 353; B. M. McCoy and T. T. Wu, *The Two-Dimensional Ising Model* (Cambridge, Mass.: Harvard University Press, 1973).

[4] Factors of $\hbar$ resp. $\hbar^2$ have been incorporated into H and J.

and the σ_i take on the values ± 1, which may be pictured simply as replacing the atoms at the lattice sites by arrows pointing either up or down. The question then is whether, with long-range order at low temperatures—all the arrows pointing in the same direction—there is a transition temperature at which this order (taken as the sign of ferromagnetism) abruptly disappears. The model proposed by Ising possesses three important ingredients: it contains enough of the physics to make it relevant; it is simple enough to make the problem posed by it amenable to rigorous attack; and the resulting mathematical problem is nontrivial, representing a serious challenge to mathematically ambitious theorists.

In the simplest case of a one-dimensional crystal, with all the atoms on a straight line, it was soon found that there is no ferromagnetism at any temperature: no matter how low the temperature, long-range order can always be destroyed by turning around all the spins on one side of a single point in the chain. However, in the two-dimensional case, Onsager (and later others, who used different mathematical methods) was able to demonstrate that there is indeed a phase transition from the ordered state at low temperature to a state without long-range order.[5] The three-dimensional case is still unsolved and may well remain so for a long time.

This vastly oversimplified toy model is the closest thing we have to an explanation of the ferromagnetic phase transition. It certainly cannot be called a *theory* of ferromagnetism. Nevertheless, the model serves to persuade us that quantum mechanics, together with thermodynamics, indeed *explains* the behavior of a ferromagnet.[6] Clearly it would be preferable to have a realistic theory of this phase transition, one that could at the same time allow us to calculate the precise value of the Curie temperature. But sometimes a model is the best we can come up with, and the fact that such models are regarded as explanations demonstrates how the use of mathematics in physics goes far beyond the mere

[5] Lars Onsager, "Crystal statistics. I. A two-dimensional model with an order-disorder transition," *Physical Review* **65** (1944), p. 117.

[6] The same Ising model has also been used for a number of other purposes, among them as a trial ground for approximation methods.

production of numbers that can be compared with experimental data.

There are many other instances in which simplified models were introduced for mathematical purposes because the equations of the realistic theory are too difficult to solve even approximately. This is particularly so in field theory. One case is known as the Schwinger model,[7] a version of quantum electrodynamics in one space dimension, with a massless Dirac matter field; another is the Thirring model,[8] which is a self-coupled spinor field in one space dimension. These are less ambitious and less challenging than the Ising model, their main virtue being that their equations can be solved. As a result, they have served for many years as proving grounds for approximation schemes or solution methods such as lattice field theories, the latter models in their own right, in which continuous space or space-time is replaced by a discrete lattice, thereby avoiding the ultraviolet (i.e., small-distance) divergencies that usually plague the continuum theories.

The role of mathematics, however, is not restricted to such analyses and applications within models; in many instances, mathematics has served as a seminal tool of thought. The use of differential geometry, first in the general theory of relativity, but now in all of classical mechanics,[9] has greatly stimulated our imagination in these areas, and the detailed properties of infinite-dimensional Hilbert spaces and linear operators on them form the backbone of quantum mechanics. Certain aspects of elementary-particle theory are completely dominated by applications of the theory of group representations. (See chapter 5 for more details.) Noether's theorem, which establishes a connection between transformational invariances of a theory and its conservation laws, represents a very general application of mathematics to symmetries in the laws,

[7]Julian Schwinger, "Gauge invariance and mass. II," *Physical Review* **128** (1962), p. 2425.
[8]Walter Thirring, "A soluble relativistic field theory," *Annals of Physics* **3** (1958), p. 91.
[9]See, for example, the treatises by Ralph Abraham and Jerrold E. Marsden, *Foundations of Mechanics* (Reading, Mass.: Benjamin/Cummings, 1978), and V. I. Arnold, *Mathematical Methods of Classical Mechanics* (New York: Springer-Verlag, 1978).

truly demonstrating the power of abstraction over ordinary intuition. (We shall discuss that, too, in more detail in chapter 5.)

During the heyday of "S matrix theory," in the 1960s, many physicists believed that underlying and determining the properties of all the elementary particles was the analytic behavior of the mother of all S matrices—the sole mathematically consistent all-encompassing function embodying the requirements of causality and relativistic invariance. Nature, in other words, was, given a few constraints, constructed in the only mathematically possible way. This theory is no longer accepted (and never was accepted by all physicists), but the slightly less ambitious aim of those who are searching for a "theory of everything" is also to lay down a simple set of extremely general principles which uniquely imply the fundamental field equations governing the universe. Whether one believes such grandiose goals will ever be reached or not, they show the stimulating effect of mathematics on the imagination of physicists. Einstein was surely right in declaring that "the creative principle [of science] resides in mathematics,"[10] as far as physics is concerned, but we wonder why? Before we can try to answer it, we're going to have to examine the nature of mathematics.

WHAT IS MATHEMATICS?

For centuries mathematicians have puzzled and quarreled over the fundamental nature of mathematical knowledge. On one hand, there is the old and well-established camp of "Platonists," which includes most of the great and creative mathematicians of the past and present. They see mathematical truths as having an independent existence in a Platonic universe, and the role of mathematicians as analogous to that of experimental physicists (but exploring the realm of eternal abstract ideas instead of the real universe of physics), *discovering* these truths when they state and prove a theorem.

---

[10] Albert Einstein, *Ideas and Opinions* (New York: Crown, 1954), p. 274.

Situated at the opposite extreme from the Platonists is the younger school of *intuitionism*, for whom mathematics is created by human intuition, a product of our mind, led by the Dutch mathematician Luitzen Brouwer and counting Hermann Weyl among its prominent members. Since the architectural structure of mathematics is erected by us, the intuitionists will allow only the use of explicitly constructive methods. For example, they will not consider any proof valid that uses indirect procedures such as the techniques of *reductio ad absurdum* or of mathematical induction. In order to prove a theorem, it is not sufficient, in their view, to use an elegant proof method employed by many mathematicians, namely to demonstrate that assuming it to be false leads to a contradiction; to be considered valid, a proof must be *constructive*. For instance, to prove the existence of a solution to an equation, in a given space of functions, they demand a procedure by which the solution can be calculated; showing that the assumption of a lack of solutions is self-contradictory would be insufficient. This school of thought was particularly influential earlier in this century, generating heated debates with the Platonists, who resented the intuitionists' strictures as impoverishing mathematics by depriving it of some of its most valuable tools. David Hilbert complained bitterly that intuitionism "seeks to break up and to disfigure mathematics" and fought strenuously to stop it.[11]

This brings us to the whole question of the need for an iron-clad *proof* if a statement is to be admitted as a mathematical theorem. Would it not be sufficient to treat a theorem like a scientific theory and allow it to be corroborated by confirming instances, especially if one can employ a computer to run through a large number of tests in a short time?[12] Or perhaps one could be utterly convinced of its truth through one's intuition, a method used by the brilliant Indian mathematician Srinivasa Ramanujan, often with impressive, beautiful, and valid results (as confirmed by later

---

[11] David Hilbert, *Gesammelte Abhandlungen* (Bronx, N.Y.: Chelsea, 1965), vol. 3, p. 159.

[12] For an amusing parable along these lines, see John D. Barrow, *Pi in the Sky: Counting, Thinking, and Being* (Oxford: Clarendon Press, 1992).

proofs). These means of substantiating theorems might seem reasonable to many scientists, especially to the less rigorously inclined theoretical physicists, who frequently employ such methods with great success.

Though Pythagoras and his followers felt the esoteric knowledge acquired in this new manner did not necessarily have to be shared with the rest of the world, the notion of requiring a mathematical theorem to be proved once and for all by a foolproof procedure whose correctness can be checked by anyone sufficiently intelligent and knowledgeable in mathematics has come down to us from the ancient Greeks. It did not occur to mathematicians in other cultures, though they discovered many extremely valuable mathematical facts. The idea of requiring a proof turned out to be enormously fruitful, not merely because it eliminated many plausible though false statements, but because the methods employed in these proofs aften became fertile sources of fresh ideas leading to new avenues of inquiry. Mathematics without proofs would have been a barren and far less active field than it has proven to be for well over two thousand years.

To be sure, what is meant by an "ironclad proof" has changed in the course of history. As the methods and ideas of mathematics expanded, so did the need for more sophisticated proof procedures. Whereas for many centuries, Euclid served as a model to be emulated—the ideal of rigorous reasoning both within and outside mathematics—on later occasions even some of the arguments used by great mathematicians were subsequently found to be inadequate. Euler's theorem for polyhedra is an excellent example. He convinced himself of his well-known formula $V - E + F = 2$ (where V is the number of vertices, E the number of edges, and F the number of faces) by examining a large number of cases and finding it always correct. An actual proof was not found until some sixty years later by the French mathematician Cauchy, who used an ingenious argument based on envisaging the polyhedron made from a rubber sheet and flattening it out after removing one face. However, it did not take long for mathematicians to find some strange kinds of counterexamples that eluded Cauchy's proof. This led to a string of further and further restrictions in the definition of what constitutes the class of polyhedra for which Euler's theorem is valid, and in this succession the argumentation became ever more rigorous. When a later counterexample shows a proof of an interesting and valuable theorem to be faulty, rarely is it discarded altogether; rather, the theorem is stated with tightened hypotheses so as to exclude the counterexamples. At the same time, the participants learn an important lesson about the class of objects to which the theorem applies, and in the process, their imagination is sometimes greatly stimulated, leading to further progress.

WHAT DOES RIGOR DO FOR PHYSICISTS?

During the nineteenth century, mathematicians paid increasingly close attention to the rigor with which they proved theorems, to the point where there should be no chance of inadvertently overlooking a possible escape from the logic of each step in a proof. While these more and more inflexible rules have contributed to the contemporary alienation between mathematicians and physicists, who often lose patience with a rigor that appears to them mortis, they also eliminated any possibility of an unconscious assumption sneaking into a proof. In this, Poincaré sees one of the most valuable contributions mathematical physics can make. The most dangerous hypotheses, he writes, are "those which are tacit and unconscious. Since we make them without knowing it, we are powerless to abandon them. Here again, then, is a service that mathematical physics can render us. By the precision that is characteristic of it, it compels us to formulate all the hypotheses that we should make without it, but unconsciously."[13]

Instances in which a lack of mathematical sophistication has led physicists to incorrect conclusions are not hard to find. Scattering theory, at the basis of which lies a comparison between the devel-

---

[13] Henri Poincaré, *The Foundations of Science*, p. 134.

opment of the state of a system of interacting particles and that of a similar system without interactions, starting out equally in the distant past, offers an example. Since e^{-iHt} is the operator describing the time development of a system whose Hamiltonian is H, and $e^{-iH_0 t}$ is the corresponding operator for the noninteracting particles, the operator of interest is $U(t) = e^{iHt} e^{-iH_0 t}$, and particularly its limit as $t \to -\infty$, known as the wave operator. Since H and H_0 are self-adjoint operators, $U(t)$ is unitary. If the Hilbert space of the system were finite dimensional—and the intuition of physicists is usually honed to operators on finite-dimensional spaces, i.e., matrices—it would follow that the wave operator, as the limit of a sequence of unitary operators, is also unitary. On an infinite-dimensional Hilbert space, however, this is not necessarily so,[14] and for many years physicists were led to incorrect and puzzling conclusions. Such difficulties, especially for systems consisting of more than two particles—say, in three-particle problems like the scattering of a particle by a bound state of two others—still sometimes bedevil theorists who find conclusions based on strict mathematics counter-intuitive and hard to accept.

In another important instance of the use of rigorous mathematics, the result is not yet known. Chapter 7 alludes to the infinities besetting quantum field theories and how they are eliminated by renormalization. In spite of the fact that QED allows us to calculate, by this procedure, numerical predictions that are confirmed by experimental data with extraordinary precision, many of the most prominent workers in field theory, including both Dirac and Feynman, have expressed their strong misgivings about sweeping nonsensical results under the rug. What do these infinities tell us about the theory? One possibility is that we are simply not smart enough to calculate the numerical implications of the equations of QED without resorting to series expansions in powers of the fine-structure constant α; if we could avoid such expansions, no infinities would ever arise. In this scenario, the solutions of the equations cannot be expanded in convergent power

[14] Which also implies that what is meant here by the *limit* has to be stated more precisely.

series, and the infinities are an indication of the lack of convergence of what may at best be asymptotic expansions. There is, however, another possibility: the equations simply *have no solutions,* and the infinities signal to us that what we are trying to approximate by the first few terms in our power series just does not exist.

The answer to this difficult mathematical question is not known, but there are indications that the second of the two possibilities mentioned may be correct. If that should turn out to be so, the interesting further question would be: how can equations that have no solutions lead to results that agree with nature so accurately? What are we approximating so well, if not a solution? It is clear that we would be faced with another mathematical problem, namely, to redefine what is meant by a *solution* of these equations, a problem that mathematicians have confronted a number of times in the past in other contexts. In order to overcome it, they have had to introduce devices such as "weak solutions" of partial differential equations. In other words, the simple statement "the equations have no solution" would not be the end of the matter, but the beginning of a new mathematical investigation.

I do not wish to overemphasize the questions of rigor and the nature of mathematical knowledge in examining the role of that knowledge in physics. Controversies about rigor and deep problems about the safety of the grounds on which the structure of mathematics is built need not, ultimately, be our concern, except for one issue: how universal can we assume this structure to be? There is no reason to fear that because of Gödel's theorem one day all of mathematics might crumble and take physics down with it.[15] (I know of no mathematician who anticipates catastrophic consequences of Gödel's important result.) That in every formal mathematical system of sufficient complexity certain statements cannot be proved to be either true or false (but may then be decidable

---

[15] Gödel's theorem states that every consistent axiomatic system strong enough to encompass arithmetic permits the formulation of statements that cannot be proved, within the system, to be either true or false; they are formally undecidable.

outside the system) is of no real interest for applications.[16] Gödel's theorem, however, does appear to weaken the Platonic view and would seem to strengthen the argument that mathematics is a human construction and therefore not necessarily universal. Aliens from another galaxy are likely to have employed other axioms and proved other theorems. This is not to say that they will deem some of our mathematical results incorrect, nor that we will find theirs erroneous—I do not believe their mathematics will *contradict* ours—but that the structures they have erected are likely to differ from ours, and so will their science.

WHAT ACCOUNTS FOR ITS POWER?

The powerful consequences of formulating the laws of physics in mathematical language and of using the tools and concepts of mathematics are far removed from the question of rigorous adherence to the strictest rules of mathematical proofs, and as I emphasized earlier, they transcend the mere application of mathematical tools for calculating numbers. It seems remarkable that so many mathematical concepts, often developed with no thought to their usefulness in physical or any other applications, concepts that mathematicians invented (or discovered) with no end in mind but to build beautiful structures with a sublime architecture, turn out to be enormously fertile for physical theories. The theory of groups, discussed earlier, is one such example; complex numbers and analytic functions are another; non-Euclidean geometry, a third; and infinite dimensional Hilbert spaces, a fourth. Group theory originated for purely algebraic purposes; complex numbers, too, arose from algebraic needs (without them the fundamental theorem of algebra—every algebraic equation of degree n

---

[16] For example, before it was finally proved, Fermat's last theorem was thought by some to be formally undecidable. If that had been the case, the theorem would have had to be true, because otherwise there would have to exist a counterexample, and finding it would have served to disprove Fermat. In other words, in this and analogous cases, a proof of undecidability would amount to an indirect proof of correctness. Not being able to decide *within a system* whether a proposition is correct does not mean it cannot be decided at all.

has precisely *n* solutions—would be false); non-Euclidean geometries were found in answer to the question of the independence of Euclid's fifth axiom from his other axioms; and the concept of Hilbert space had its origin in expansions of functions on the basis of orthogonal polynomials. Without these ideas, and many more of a similar nature, modern physics would be unthinkable.

How do we account for this? That is the question Wigner posed in puzzlement, but did not answer, in his famous essay entitled "The Unreasonable Effectiveness of Mathematics in the Natural Sciences"; he called it a miracle. The question would seem, at first sight, to be especially vexing for Platonists, who have to construct an indeed miraculous connection between most of the nooks and crannies of the Platonic universe of ideas and the structure of the physical world. But then, if the real world is merely a corruption of ideal essences, and these are, at their core, of mathematical origin, the riddle of the connection might be solved. Those who believe mathematics is a creation of the human mind, developed in the course of evolution and hence influenced by nature, have another way of accounting for it. Some of them find no great miracle in the fact that the human brain, as it evolved in the course of eons, should be adapted so as to generate concepts that would be useful to the description of the very nature that influenced it during its development.[17]

As a physicist I do not find the Platonist view congenial, but I also cannot regard this last argument as persuasive. The most elementary mathematical structures—numbers and the elements of geometry, for example—can easily be imagined to have been imprinted on the brain in the course of evolution. They are the stuff of every-day perceptions, necessary to orient ourselves in the world. But it is hard to imagine how concepts should have adaptively evolved that are totally useless until we try to understand nature at levels enormously remote from everyday experience, to understand the world under conditions of gravity much stronger

[17] This is, for example, Michael Atiyah's position as he stated it at a panel discussion during the Eleventh International Congress of Mathematical Physics, Paris, 1994.

than on Earth or at scales far too small for our senses to take in. We'll leave this question for the time being to look, instead, at why mathematics plays such a powerful role in physics.

Some physicists essentially subscribe to Galileo's contention that "mathematics is the language of nature," paraphrased elsewhere as "God is a mathematician." In other words, mathematics is built into the very structure of nature; if we want to understand the universe, we have to learn its language and the principles of its construction. I believe this view to be mistaken: nature just *is*; it speaks no language and follows no plan; language and plans are human additions.

In my view, mathematics is an enormously powerful outgrowth of logical thought,[18] and therefore an instrument that allows us to come to conclusions we could either reach without, but much more arduously, or not at all. "The power of mathematics," Ernst Mach observes, "rests on its evasion of all unnecessary thought and on its wonderful saving of mental operations,"[19] Group theory is a good example of this economy. We could undoubtedly arrive at all the results obtained by applications of group theory without its use; however, it would be a much more difficult and lengthy process. Many physical theories and their results were, in the past, stated in a cumbersome, laborious, and sometimes ambiguous language we have difficulties understanding now. Even at a superficial level, anyone who has ever had to struggle with a clumsy way of writing equations can attest to the great intellectual facility afforded by an elegant notation. The increasing mathematization of physics has allowed us to formulate and handle the same theories and results much more clearly, efficiently, and unambiguously.

This, then, is the real role and power of mathematics: an efficient tool of sharpened logical thought. If, from premises we are

[18] I am treading here on gound that is very controversial among mathematical logicians: the *logicists* hold that mathematics is nothing more than a part of logic; others regard mathematical objects as separate and beyond logic.
[19] Quoted on p. 106 of Freeman Dyson, "Mathematics in the physical sciences," in *The Mathematical Sciences*, ed. COSRIM (Cambridge, Mass.: MIT Press, 1969), pp. 97–115.

willing to grant on the basis of experimental evidence, mathematicians have already spent much labor and ingenuity on arriving at certain far-reaching conclusions without further effort on our part, there is every reason to make use of these handy results and shortcuts. Mathematical physicists develop these shortcuts themselves, if necessary, for immediate and later use in other contexts.

What is more, the use of such efficient tools not only saves us time and effort, it frees us to use our imagination for the more important task of creating scientific structures with new ideas. This is presumably what Einstein meant when he declared "the creative principle [of science] resides in mathematics." A fuller expression was given by Freeman Dyson: "A physicist builds theories with mathematical materials, because the mathematics enables him to imagine more than he can clearly think.... In the process of theory building, mathematical intuition is indispensable because the 'evasion of unnecessary thought' gives freedom to the imagination."[20] Is it any wonder, after finding mathematics such a marvelously useful tool, that we tend to reach for it at the least opportunity? Here is a prefabricated architecture—constructed in part with our purposes in mind—which we would be foolish not to employ as essential structural building material for our science. However, it is also well to attend to Dyson's warning, which follows: "Mathematical intuition is dangerous, because many situations in science demand for their understanding not the evasion of thought but thought," a reminder that, after all, there is a difference between the aims of physicists and the goals of mathematicians.

What appears miraculous is that the razor-sharp blade devised by mathematician-artists sometimes only for the sake of beauty turns out to be the most efficient instrument for understanding nature, and that so many of its varieties, elegantly created as *l'art pour l'art*, are found to come in handy. The premise, however, that mathematicians develop these refinements of logical thinking with no attention to their utility is somewhat of an exaggeration, and so is the allegation that even the most esoteric parts of mathematics eventually find their way into our science. Numerous results

[20] Ibid.

of mathematics are of no interest to physicists. Furthermore, even at a time when many mathematicians pride themselves in taking no interest in the applications of their work, once an area becomes useful, it is stimulated to grow in the direction from which the sun of applications shines. And since one of the cardinal virtues of mathematics is its great generality, it should come as no surprise that its results lend themselves to a wide range of different applications, including the needs of physics. Moreover, since many areas of mathematics, especially in earlier days, originated with problems posed by physics, perhaps some of the mystery is not so mysterious, after all. Nevertheless, the symbiosis of physics and mathematics is the greatest marvel in the structure of science, and each would be immeasurably poorer without it.

Mathematics, then, is essential for physics, not because it is embedded in the skeleton of nature itself, but because we need this powerful and efficient tool for our thought processes; without it our thinking would be stymied. That "nature is mathematical" is not a miracle: we do not *find* mathematics in nature; our thinking puts it there. Our job as physicists is to understand nature and to use the most efficient tools and language for this purpose, not to explain why the tools and language exist.

APPENDIX ON SOLITONS

The history of the Korteweg–de Vries (KdV) equation goes back to the observation, in 1834, by John Scott Russell of heaps of water produced by ships in the Union Canal in Edinburgh, which he followed, in astonishment, on horseback for miles. Preserving its shape for a long time, the velocity of the "great solitary wave" seemed to depend on its size, and he challenged "future wave mathematicians" to come up with an explanation of "this singular and beautiful phenomenon." Such an explanation, however, did not appear until some sixty years later, when the Dutch mathematicians D. Korteweg and G. deVries were able to use hydrodynamics to derive the following equation for the one-dimensional motion of a wave on the surface of a shallow water channel:

$$\frac{\partial u}{\partial t} = \frac{3}{2}\sqrt{\frac{g}{h}}\frac{\partial}{\partial x}\left(\frac{1}{2}u^2 + \frac{1}{3}\sigma\frac{\partial^2 u}{\partial x^2}\right),$$

where h is the depth of the water, g the acceleration of gravity, $\sigma = \frac{1}{3}h^3 - \frac{Th}{\rho g}$, ρ the water density, and T the surface tension. A simple change of scale brings this *KdV equation* into the form

$$\frac{\partial u}{\partial t} = 6u\frac{\partial u}{\partial x} + \frac{\partial^3 u}{\partial x^3}, \tag{1}$$

which is now known to be applicable to many other physical phenomena, such as collision-free hydromagnetic waves, acoustic waves in anharmonic crystal lattices, ion-acoustic waves in cold plasmas, pressure waves in liquid-gas bubble mixtures, waves in elastic rods, and others.

Equation (1), being nonlinear, is not subject to the superposition principle, and its wavelike solutions behave quite differently from those of the linear wave equation.[21] There are, to begin with, "traveling waves" of the form $u(x, t) = f(x + ct)$, but in contrast to traveling solutions of the wave equation, these have to have a very specific shape (see fig. 3.1), namely

$$u(x, t) = \frac{c}{2}\left/ \cosh^2\left[\frac{1}{2}\sqrt{c}(x - x_0 + ct)\right].\right. \tag{2}$$

This is Scott Russell's "great solitary wave" (whose amplitude is $c/2$ and whose speed is c, where c is an arbitrary constant)[22] moving to the left, and tending to zero as $x \to \pm\infty$ for fixed t.

For many years the solitary waves were the only solutions known until, in 1965, Gardner, Greene, Kruskal, and Miura decided to find solutions by computer. Specifically, they asked themselves what would happen to two solitary waves of different speeds that came in from infinity on the right, with the slower one in front. So long as the two are far apart, their product in

[21] Since it is odd in x and t, it has a unidirectional character, but this is of no great significance.

[22] Note that, in contrast to the wave equation, the propagation speed of wave solutions here is not contained in the equation but is arbitrary.

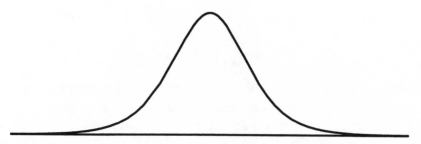

Figure 3.1. A solitary wave solution of the KdV equation.

the quadratic term will be negligible, the superposition principle will hold to a good approximation, and they will retain their shape. However, the further development found by means of the computer came as a great surprise. As the two "heaps" approach each other, their shapes become distorted by the nonlinearity in the equation, as can be expected, but eventually the two separate again, now with the faster one in front, their shapes and sizes just as they had been initially, with the only evidence of their past interaction a shift in position, one forward and the other backward, from where they would have been in the absence of the other. (The same phenomenon was observed for more than two initially solitary waves.) It was as though the incoming "lumps" had some kind of identity they retained during and after their mutual interaction; these particle-like solutions Kruskal and Norman Zabusky named *solitons.*[23]

Remarkable as these newly discovered solutions of the KdV equation and other nonlinear differential equations were, the fact that they were seen on computers did not lead to an understanding of why they occurred: the computer printouts did not *explain* them. That explanation was found a short time later and was just as surprising as the discovery of the solitons themselves.[24] Here is an outline of the explanation, as generalized by Peter Lax.[25]

[23] N. J. Zabusky and M. D. Kruskal, "Interaction of 'solitons' in a collisionless plasma and the recurrence of initial states," *Physical Review Letters* **15** (1965), p. 240.
[24] C. S. Gardner et al., "Method for solving the Korteweg–de Vries equation," *Physical Review Letters* **19** (1967), p. 1095.
[25] P. D. Lax, "Integrals of nonlinear equations of evolution and solitary waves," *Communications in Pure and Applied Mathematics* **21** (1968), p. 467.

Consider the time-independent Schrödinger equation in one dimension (in suitable units), with an attractive potential $-V(x)$ that vanishes as $|x| \to \infty$,

$$H\psi = \left[-\frac{d^2}{dx^2} - V(x) \right]\psi = E\psi.$$

It will have a spectrum consisting of the continuum $E \geq 0$ and a certain number of negative point eigenvalues. If V depends on a parameter s, the point eigenvalues will generally change with s. We now pose the *isospectral problem*: what kind of dependence of V on s would assure that the spectrum remains invariant as s is changed?

The answer to this question can, in principle, be given very easily: if there exists an operator A such that $\partial H/\partial s = [A, H]$, then the spectrum is invariant under a shift in s. This is because, if ψ is an eigenstate of H with the eigenvalue E, then $(\psi, H\psi)/(\psi, \psi) = E$ and

$$\frac{dE}{ds} = \left[\left(\frac{\partial \psi}{\partial s}, H\psi \right) + \left(\psi, H\frac{\partial \psi}{\partial s} \right) \right] \Big/ (\psi, \psi)$$

$$- (\psi, H\psi)\left[\left(\frac{\partial \psi}{\partial s}, \psi \right) + \left(\psi, \frac{\partial \psi}{\partial s} \right) \right] \Big/ (\psi, \psi)^2$$

$$+ \left(\psi, \frac{\partial H}{\partial s}\psi \right) \Big/ (\psi, \psi)$$

$$= (\psi, (AH - HA)\psi)/(\psi, \psi) = (\psi, (AE - EA)\psi)/(\psi, \psi) = 0.$$

The trick now is to find an operator A of such a form that it is only the potential $V(x)$ that changes with s and not the kinetic-energy term $-\frac{d^2}{dx^2}$. The simplest form for such an A is $A = c\frac{d}{dx}$, in which case we obtain $\frac{\partial V}{\partial s} = c\frac{\partial V}{\partial x}$, and the solution of this equation is clearly $V(x, s) = V(x + cs)$, a result that is not surprising; a simple shift of the potential, of course, leaves the eigenvalues unchanged.

The next simplest choice that works is $A = 4\frac{\partial^3}{\partial x^3} + 6V\frac{\partial}{\partial x} + 3\frac{\partial V}{\partial x}$; this leads to the partial differential equation

$$\frac{\partial V}{\partial s} = 6V\frac{\partial V}{\partial x} + \frac{\partial^3 V}{\partial x^3},$$

which is identical with (1) if the parameter s is identified with the time t in that equation (not the time in the Schrödinger equation!). In other words, if the potential $-V$ in the one-dimensional Schrödinger equation depends on a parameter t in such a way that it satisfies the KdV equation, then its bound state eigenvalues remain unchanged. Furthermore, we can also calculate the asymptotic form of the bound-state wave functions as functions of t, and we find that

$$\psi_{\text{bd}}^{(n)} \sim \gamma_n(t)e^{-\kappa_n|x|},$$

where

$$\gamma_n(t) = \gamma_n e^{4\kappa_n^3 t}$$

is the "norming constant" and $-\kappa_n^2$ the energy of the nth bound state. Similarly, the reflection coefficient for particles incident on this potential changes in the following simple way:

$$R(k, t) = R(k)e^{-8ik^3 t}.$$

At this point we use the important fact that there is a straightforward method for solving the "inverse problem" of determining the potential in the Schrödinger equation from a knowledge of these three sets of data: the reflection coefficient R for all k, all the bound state eigenvalues $-\kappa_n^2$, $n = 1, 2, \ldots$, and the norming constants γ_n for each of them. This is accomplished by solving a linear integral equation of the Fredholm type, the Marchenko equation.[26]

There is, therefore, a remarkable way of solving the initial-value problem of the (nonlinear) KdV equation by *linear* methods: (1) using the initial value $u(x, 0)$ as the potential $V = -u$ in the Schrödinger equation, calculate the reflection coefficient $R(k)$ for

---

[26] See, e.g., K. Chadan and P. C. Sabatier, *Inverse Problems in Quantum Scattering Theory* (New York: Springer-Verlag, 1989).

all k, the bound-state eigenvalues $-\kappa_n^2$, and the corresponding norming constants γ_n; (2) using $R(k, t)$, κ_n^2, and $\gamma_n(t)$ as input for constructing the kernel of the Marchenko equation, solve that integral equation to find the potential $V(x, t)$. The result $u(x, t) = -V(x, t)$ then is the solution of the KdV equation at the time t with the given initial values.

Now, where do the solitons come in? Suppose you assume $R(k) \equiv 0$, so that you are looking for a reflectionless (transparent) potential in the Schrödinger equation. The kernel of the Marchenko equation in that case becomes separable, and the equation can easily be solved in closed form. If there is but one bound state of energy $-\kappa^2$, the potential is found to be

$$V(x) = -\frac{2\kappa^2}{\cosh^2[\kappa(x + x_0)]},$$

where x_0 depends on the norming constant as $x_0 = \log(2\gamma^2\kappa)/2\kappa$ and therefore depends on t as $x_0 = 4\kappa^2 t + x_{00}$. The function $u(x, t) = -V$ is thus precisely of the form (2) with $c = 4\kappa^2$. In other words, the negative of a transparent potential with one bound state of energy $-\kappa^2$ is the solitary-wave solution of the KdV equation of speed $4\kappa^2$ and amplitude $2\kappa^2$. Similarly, the negative of a transparent potential with N bound states of energies $-\kappa_n^2$, $n = 1, \ldots, N$, is the N-soliton solution of the KdV equation with soliton-speeds $c_n = 4\kappa_n^2$.

The general initial value problem for the KdV equation cannot, of course, be solved in closed form, but it can always be solved via the linear Marchenko equation and this allows us to understand what happens in the limits as $t \to \pm\infty$. Suppose the potential $V(x) = -u(x, 0)$ in the Schrödinger equation produces N bound states of energies $-E_1, -E_2, \ldots, -E_N$. Then both as $t \to +\infty$ and as $t \to -\infty$, $u(x, t)$ becomes a sum of solitons moving with velocities $4E_1, 4E_2, \ldots, 4E_N$, plus some additional waves that, like transients, gradually disappear.[27]

---

[27]Note that a negative potential in one dimension always produces at least one bound state. So if $u(x, 0)$ is everywhere positive, it will produce at least one soliton.

An interesting question concerns the *conservation laws* associated with the KdV equation, that is, any law of the form

$$\frac{\partial \rho}{\partial t} + \frac{\partial j}{\partial x} = 0,$$

where ρ and j are functions of a solution u of (1) and its derivatives, which implies, if ρ and j vanish sufficiently rapidly at $\pm\infty$, that $\int_{-\infty}^{+\infty} \rho\, dx$ is a constant of the motion. One such conservation law is immediately obvious from (1) by writing it in the form

$$\frac{\partial u}{\partial t} + \frac{\partial}{\partial x}\left(-3u^2 - \frac{\partial^2 u}{\partial x^2} \right) = 0,$$

so that $\rho = u$ and $j = -3u^2 - \frac{\partial^2 u}{\partial x^2}$. A second one is obtained by multiplying (1) by u, which leads to

$$\rho = u^2, \quad j = 4u^3 + 2u\frac{\partial^2 u}{\partial x^2} - \left(\frac{\partial u}{\partial x}\right)^2.$$

Therefore, both $\int_{-\infty}^{+\infty} u\, dx$ and $\int_{-\infty}^{+\infty} u^2\, dx$ are constants of the motion. A third conserved quantity is the energy, the constant value of the Hamiltonian[28] $H = u^3 - \frac{1}{2}(\frac{\partial u}{\partial x})^2$ for the function ϕ which is such that $u = \frac{\partial \phi}{\partial x}$ and the equation of which can be obtained from a Lagrangian.

It turns out that the KdV equation has, in fact, *infinitely many* independent conservation laws, in all of which ρ and j are polynomials of u and its x-derivatives, which is why the equation is called *integrable*, and this property is thought to be connected to the existence of its soliton solutions.

There are now a large number of known integrable nonlinear evolution equations in one dimension that have solitonlike solutions and that can be solved by associating them with scattering solutions of a linear differential equation, using what has become known as the *inverse scattering method*.

---

[28] This is the Hamiltonian of a dynamical system whose equation of motion is the KdV equation; H was the Hamiltonian of the Schrödinger equation.

The method and the existence of solitons have been generalized, with limited success, to two dimensions, but never to higher dimensions than two. Whether there are nonlinear evolution equations in three dimensions that have "lumplike" solutions with solitonlike properties is unknown. Many physicists and mathematicians have tried to find such equations without success.[29]

[29] I have generalized the Marchenko method to three dimensions for the specific purpose of finding three-dimensional solitons, but without success as well; see R. G. Newton, *Inverse Schrödinger Scattering in Three Dimensions* (New York: Springer-Verlag, 1989).

Fields and Particles

ONE OF MICHAEL FARADAY'S greatest contributions to physics was the introduction of the notion of the field. To appreciate the scope of this concept, recall the discussion in chapter 2 of Newton's seminal approach to the problem of how particle A in position P_A influences the motion of particle B in position P_B; he broke it into two parts: (1) what force does particle A exert on particle B? and (2) what influence does this force have on the motion of particle B? The answer to the first of these questions introduced the repugnant idea of action at a distance—a force exerted by A at P_A on B at the distant point P_B. For electric and magnetic influences, Faraday, in effect, subdivided the first question into two further parts: (1a) what field does A produce everywhere in space? and (1b) what force does the field at P_B exert on particle B at the very point P_B? By doing so, he circumvented altogether the need for a force acting at a distance. This remarkable idea was then generalized and has pervaded all of fundamental physics ever since.

Faraday, who had little interest in mathematics and did not invent the field concept for mathematical purposes, imagined electric and magnetic fields in very concrete and physical terms as lines that stretched through space, exerting forces like rubber bands. Maxwell subsequently replaced these images by intricate mechanical models of interlocking wheels making up the luminiferous ether that carried light and other electromagnetic waves, but from the time Hertz opened his treatise *Electric Waves* by declaring that "Maxwell's theory is Maxwell's system of equations," we have given up all such pictorial representations and think of fields simply as abstract *conditions of space*. The notion of an all-pervading ether finally fell victim to Einstein's theory of relativity.[1]

---

[1] In a provocative article entitled "The persistence of the ether" (*Physics Today*, January 1999, pp. 11 to 12), Frank Wilczek argues that the field is today's ether,

Is the Field Real?

When we say a field is "a condition of space," what exactly do we mean? In the general theory of relativity, dealing as it does with gravitation only, the condition may be thought of as a local geometry, with the field as the metric tensor. Just as we can imagine space having a certain quality that determines its geometry—triangles with straight sides, for instance, will have interior angles that add up to either more or less than 180°—we have to accept the notion that space can have a great variety of other conditions, some of which require rather abstract mathematical representations.

For Newtonian gravity, things are still relatively simple. Since the force exerted on a massive particle has a direction as well as a magnitude, the Newtonian gravitational field has to be a vector field. Mathematically this is expressed by saying the field is a "vector-valued function" of three-dimensional space as well as time. Similarly for the electric and magnetic fields; since these, however, are tied together by the Maxwell equations in a way that is consonant with Lorentz covariance, they have to be combined into an antisymmetric (space-time) tensor of rank two: the electromagnetic field is a "tensor-valued function" of four-dimensional space-time.[2] The metric tensor of general relativity represents another tensor of second rank; it too is a tensor-valued function. So already at the classical level, the total number of components required to characterize the condition of space at a point in space-time is growing fairly large. We will discuss the quantum field later on.

---

and we cannot do without it. While this is an interesting point of view, it ignores some important differences: the new, much more abstract "ether" certainly has none of the corporeal qualities of the one it replaced, and it cannot serve as a frame of reference. To call the field the "ether" is really no more than a matter of language.

[2] "Vector-valued" and "tensor-valued" functions of space-time are simply vectors or tensors whose components are functions of space-time. For a review of vector and tensor functions, see, for example, J. D. Jackson, *Classical Electrodynamics* (New York: John Wiley & Sons, 1975). See also chapter 5.

One might conclude from the manner of Faraday's introduction of the electric and magnetic fields that, notwithstanding his very physical notions of rubber-band-like lines of force, which we no longer take seriously, the fields might be regarded simply as auxiliary concepts with little claim to reality. Such a view, however, would have been hard to maintain after the formulation of Maxwell's equations and their implication that the fields carried both energy and momentum. The Maxwell-Lorentz equations, after all, imply that moving electric charges lose kinetic energy and momentum during their motion—losses that can be accounted for by viewing them as stored in the field—and they also pick up kinetic energy and momentum from the field. Furthermore, electromagnetic waves transport both energy and momentum over long distances—light exerts pressure—and the transport takes time to travel at the speed of light. Only the total energy and the total momentum of the charged particles and the field together are conserved. There is, therefore, good reason to ascribe reality to the field as a condition of space. It has to be considered just as real as the charges themselves—perhaps even more so.

Differential Equations for the Fields

If Faraday's principal aim in introducing electric and magnetic fields was to rid physics of action at a distance, the field concept alone achieved this purpose only partially. After all, there is little conceptual difference between stating Coulomb's law directly for the force between two distant charges—the force is inversely proportional to the square of the distance between the charges—and formulating the same law for the field at a distant point P_B, where the field strength is inversely proportional to the square of the distance of P_B to the charge at P_A. Maxwell's achievement was carrying the aim of abolishing action at a distance to its logical conclusion by formulating the laws governing the electromagnetic field in terms of partial differential equations. While it is true that Coulomb's law is nothing but the solution of Poisson's equation (with appropriate boundary conditions), there is a profound conceptual difference between the two. The former expresses the field at point P_B directly in terms of the charges at points P_{Ai}, $i = 1, 2, \ldots$, and their distances from P_B, as would be proper for a force in an action-at-a-distance theory, whereas Poisson's partial differential equation *relates the value of the electric field at each point to its values at infinitesimally neighboring points.* The physical influences are thus carried not directly from the charges to distant points in space but continuously from one point to another contiguous to it. Stating the laws governing the field in the form of partial differential equations, a procedure followed ever since Maxwell, is therefore the logical outcome of abolishing action at a distance.[3] In many specific cases, in order to solve them or to study their solutions, it is convenient to replace the differential equations, together with their boundary conditions, by integral equations, but the original differential form is the physically most appropriate manner of stating the laws.

The same arguments apply to the gravitational field, for which Einstein's partial differential equations of the general theory of relativity superseded Newton's original action-at-distance formulation. The only difference is that in the case of general relativity, Einstein chose the option of viewing the condition of space as geometry rather than as a field, which is primarily a matter of language, a language that may or may not turn out to be appropriate for quantum gravity.[4] We must also recognize that any attempt to describe this condition of space independently of the state of motion of the reference frame from which it is viewed has to be abandoned. Maxwell's equations for the electromagnetic field were the first to force us to that conclusion; Einstein's theories—both the special and the general—formalized it for all theories.

---

[3] If it should turn out that space-time has an inherently grainy structure, difference equations may have to replace differential equations, but these would achieve the same end result.

[4] To say that the geometrical description of gravity is primarily a matter of language is not to deny that it has empirical content; after all, this language works for gravity but not for electromagnetism.

QUANTUM FIELDS

Of course we now know that Maxwell's description of the electro-magnetic field and Einstein's of gravity are inadequate and have to be superseded by quantum descriptions. No one has as yet succeeded in this task for the gravitational field, but QED is a highly successful replacement of the classical Maxwell-Lorentz theory for electrons interacting with electromagnetic radiation. Additional quantum field theories have been formulated for other interactions: quantum chromodynamics (QCD) for the strong and electroweak for the weak. What all of them have in common is that they assign an *operator* rather than a set of numbers to be the condition of space.[5] In other words, the quantum fields have to be "operator-valued functions" of space-time rather than functions taking on numerical values.[6] That these operator-valued fields are not necessarily directly physically observable or visualizable is no serious objection, for there are many theoretical concepts in physics that lack direct observability or measurability. Indeed, even numerical vector-valued, tensor-valued or spinor-valued conditions of space are difficult to visualize, since their numerical values depend on the orientation and state of motion of the reference frame. The visualizability of operator-valued fields differs from these only in degree.

---

[5] In addition, these operators must transform like scalars, vectors, tensors, or spinors under Lorentz transformations, so that the field equations have the same form in all coordinate frames.

[6] I disagree with the contrary argument put forward by Paul Teller, *An Interpretive Introduction to Quantum Field Theory* (Princeton: Princeton University Press, 1995). The word "values" does not have to be interpreted as "numerical values." A "real-valued function" maps points into a subset of the real line; a "complex-valued function" maps them into the complex plane; an "operator-valued function" maps them into a ring of operators on a specified Hilbert space. However, we have to add one caveat: it turns out to be mathematically inconsistent to assume the fields to be pointwise defined. Instead, they have to be averaged over small regions surrounding each point in space-time and, in effect, specified on smeared-out points. For the purpose of our discussion, we will ignore this mathematical detail and idealize the fields as operator-valued functions defined at each point.

As far as the "matter field" for electrons and the electromagnetic field are concerned, the old equations—the Maxwell equations for the radiation field with electrons as sources, and the Dirac equation for the electron field in the presence of electromagnetic potentials, equations which are, because of the couplings, nonlinear—are still valid in the quantum-field context, but their solutions are now assumed to be operator-valued rather than complex-valued functions; for the Maxwell equations, this is their first contact with quantum physics, but for the Dirac equation, which even in its numerical form was meant to be an equation describing quantum physics, this is *second quantization*. The first and perhaps most remarkable consequence of this change of the mathematical nature of the solutions is to lead to the appearance of *particles*. Dirac, of course, originally set up his equation on the assumption that there were particles—electrons—and his equation was to be satisfied by their quantum-mechanical state vector; in its second quantized form, no such assumption is needed. The Dirac equation is now simply the partial differential equation governing another quantum field, just as the Maxwell equations govern the electromagnetic field. What is more, the addition of appropriate commutation relations for the fields automatically yields the existence of photons of spin 1 obeying Bose-Einstein statistics and of electrons as well as positrons of spin $1/2$ obeying Fermi-Dirac statistics. Let's remind ourselves how this comes about.

EMERGENCE OF PARTICLES

As a first step, the interaction between the two kinds of fields is turned off by setting the electric charge e, and thus the dimensionless coupling parameter $\alpha = e^2/c\hbar$, equal to zero, thereby uncoupling the equations and making them linear. Next, the equations are subjected to spatial Fourier transformation. At this point, each Fourier component of the vector potential is required by Maxwell's equations to satisfy the same equation as a simple pendulum—the Fourier decomposition sets the fields up as infinite sums (or integrals) of harmonic oscillators, one for each allowed wave vector

k, whose magnitude determines the frequency $\omega = c|\mathbf{k}|$. For the radiation field, the quantization is then nothing but the quantum treatment of each of these oscillators.

Now, as you may recall, the spectrum of a harmonic oscillator of frequency ω has an especially simple structure: the energy of the ground state is $\frac{1}{2}\hbar\omega$, and all the other levels sit above it with equal spacing $\hbar\omega$ between them.[7] Apart from the "vacuum energy" $\frac{1}{2}\hbar\omega$,[8] the energy of the oscillator in any of its allowed states therefore readily lends itself to the interpretation that it consists of a number of independent *photons*, each with the energy $\hbar\omega$—when in its nth-energy state, the oscillator simply contains n photons of equal energy—an interpretation that is possible only because of the equally spaced nature of the oscillator spectrum. Accordingly, the operator (which turns out to be the negative-frequency part of the Fourier transform of the vector potential) that leads from each one of the oscillator's energy eigenstates to the next higher one may be viewed as "creating a particle" of energy $\hbar\omega$, which the operator leading one step down (the positive-frequency part of the Fourier-transformed vector potential) "destroys." What is more, the behavior of the states under spatial translations leads to the conclusion that the momentum—the generator of translations—of this particle is $\hbar\mathbf{k}$. The state vector of n photons of momentum **k** and helicity μ (left- or right-handed circular polarization) and m photons of momentum **k′** and helicity μ' is therefore (apart from a normalization factor) given by $\Phi_{nm,\,\mu\mu'}(\mathbf{k},\mathbf{k}') = [a_\mu^\dagger(\mathbf{k})]^n[a_{\mu'}^\dagger(\mathbf{k}')]^m\Phi_0$, where $a_\mu^\dagger(\mathbf{k})$ is the creation operator of a photon of momentum **k** and helicity μ, and Φ_0 is the vacuum state (no photons).

The quantum treatment of the free radiation field thus automatically leads to photons. That every such field gives rise to quanta is the result of two crucial circumstances: (1) the mathematical

---

[7] See the appendix to this chapter for more details on the quantization of harmonic oscillators.

[8] The vacuum energy gives rise to one of the infamous infinities, since the frequencies are unbounded. This infinity is easily removed by defining the energy of the physical vacuum to be zero. The technicalities of how this is to be done need not concern us here.

fact that the field may always be Fourier analyzed and, as a result of the Maxwell equations, be written as a superposition of infinitely many simple linear harmonic oscillators of different frequencies; and (2) the special structure of the quantum-mechanical spectrum of these oscillators for each Fourier component.[9] Since no other system has an energy spectrum with equal spacing, no other resolution of the field into a superposition of simple components would lead to a particle interpretation after quantization. That the photon has spin 1 is a consequence of the fact that the electric and magnetic fields behave as vector fields under spatial rotations, which quantum mechanically leads to an intrinsic angular momentum of $1\hbar$ associated with each oscillator excitation.[10] Finally, the Bose-Einstein statistics of photons—the spin-statistics theorem, which ties the BE statistics of the photon to its integral spin—is a result of the choice of commutation relations rather than anticommutation relations for the fields, and hence for the oscillator variables; this choice (if these two alternatives are, for reasons of simplicity, the only ones admitted) is dictated by requirements of consistency and relativistic causality, which we shall discuss in more detail in chapter 6.

The particulate nature of electrons emerges similarly from the Dirac equation, regarded as a partial differential equation for a spinor field (its second quantized version). The Fourier component corresponding to the wave vector $\mathbf{k}$ must obey an oscillator-like

---

[9] If the harmonic oscillators were replaced by systems with unequal spectral spacing it would still be possible to define a raising operator that takes a state up one step on the spectral ladder, but the additional energy of this state would then depend on where you started. Therefore, if one attempted to adopt a particle interpretation, the mass of the new "particle" would depend on what other "particles" were already there, and you could not add another "particle" of the kind already in existence. Furthermore, you could not destroy a particle of one kind before destroying those above it in the hierarchy, so to speak. Such a situation would make it difficult to view these entities as particles.

[10] In contrast to particles of spin 1 and non-zero mass, the photon, however, has only two possible spin states instead of three, parallel or antiparallel to its momentum, i.e., right or left-handed circular polarization. This is the result of the gauge invariance of the Maxwell equations, which leads to the loss of one degree of freedom, the one corresponding to the longitudinal field. The latter does not contribute to the radiation field and remains unquantized.

93

equation, except that consistency and relativistic causality require the spinor field operators (whose behavior under rotations leads to excitations that have an intrinsic spin of $\frac{1}{2}\hbar$) to obey *anti*commutation relations (again, the spin-statistics theorem). This has the consequence[11] that each "oscillator" of frequency ω and given spin projection has only two possible states, the ground state and the excited state of energy $\hbar\omega = \sqrt{m^2c^4 + \hbar^2|\mathbf{k}|^2c^2}$ above it, which is naturally interpreted as either vacuum—no particle—or one particle of energy $\hbar\omega$, mass m, and momentum $\hbar\mathbf{k}$. That no configuration can be excited beyond the first level is then equivalent to forbidding two particles to be in the same state—electrons obey the Pauli exclusion principle and are subject to Fermi-Dirac statistics.

It is important to notice that the "particles" emerging from the quantum field have all the principal characteristics associated with quantum particles: they lack both identity and what might be regarded as substance. Since they are nothing but states of field oscillators, it makes no sense to attach individual labels to them or to distinguish between them. Furthermore, being associated with precisely specified energies and momenta, they have, *ab initio*, no position in any meaningful sense of the word. Localizable entities must be formed out of superpositions of fields of infinitely many wave numbers, thus consisting of one-particle states of different momenta.[12] In addition, the particles that have emerged here automatically satisfy, depending on their spin—integral or half-integral multiples of $\hbar$, as the case may be—either Bose-Einstein or Fermi-Dirac rather than Maxwell-Boltzmann statistics, a connection that is one of the most important results achieved by quantum field theory.[13] *Whatever the quanta of the field are, they are not simply miniature billiard balls.*

---

[11] See the appendix to this chapter.

[12] How to do this was shown by T. D. Newton and E. P. Wigner in "Localized states for elementary systems," *Reviews of Modern Physics* **21** (1949), pp. 400–406.

[13] For a good review of the spin-statistics theorem, see Ian Duck and E.C.G. Sudarshan, "Toward an understanding of the spin-statistics theorem," *American Journal of Physics* **66** (1998), pp. 284–303.

If the discrete entities emerging from quantum fields lack some of the qualities we intuitively associate with "particles" at the macroscopic level, they do possess one property that no classical mechanism could confer upon them: their absolute stability and interchangeability. Tiny billiard balls could not only be broken but they could have small nicks and blemishes, making them unsuitable as ultimate building blocks of matter or radiant energy. No classical theory could account for their perfection; it takes quantum field theory to give quanta the most essential characteristics required to be called *ultimate*. Lack of "substance" is a small price for that.

In some areas, such as solid-state and nuclear physics, and in other "effective field theories," particle-like objects emerge out of quantized vibrations in a manner entirely analogous to the appearance of photons from the quantized oscillations of the electromagnetic field. These are the phonons and other entities usually referred to as *quasi-particles*. Are they any less real than photons or electrons? Or perhaps we should ask, are mesons and neutrinos any more real than phonons? Adding the prefix *quasi* to their name certainly indicates that even the physicists working closely with them tend to see them as less particle-like. When one group of discrete entities arises out of the quantization of field oscillations in vacuum and the other group by the same mechanism out of mechanical oscillations in a many-particle system, is that difference a good physical reason to accept the former and to spurn the latter, or should we call such a distinction metaphysical? One might well argue that at the submicroscopic level all particles should be treated as equally real or unreal, none more so than others.

INTERACTIONS

The above discussion was predicated on the preliminary assumption that there is no interaction between electrons and the radiation field. When we "turn on" the coupling between the electromagnetic and the electron "matter" field—taking the charge e to be nonzero—things become more complicated. The first effect is a bonus: whereas the original Dirac equation possessed solutions of positive and negative energy, its second-quantized version admits only positive energies (even though there are negative frequencies), thereby obviating Dirac's ingenious but rather artificial introduction of the infinite sea of negative-energy electrons; instead, when we allow e to be nonzero, it leads to particle solutions of positive and negative *charge*, and we automatically obtain both electrons and positrons from the quantum field (without the need for holes in a Dirac sea).

The other effects are more subtle. Since the field does not conserve the number of particles—the only conserved quantities are total energy, momentum, angular momentum, and charge—the theory predicts the possible creation of electron-positron pairs by photons, and the emission of photons by electrons, called *bremsstrahlung*. (The emission of radiation by accelerated charges is, of course, already implied by the classical Maxwell-Lorentz theory; in both theories this is the result of their nonlinearity.) These phenomena are predicted to occur not only "really" when sufficient energy is available,[14] but also "virtually" all the time in the sense that perturbation theory, which has to be employed for the calculation of every physical process, requires their presence in "intermediate states."

The necessary appearance of such virtual processes produces a number of physical effects. Owing to the constant presence of a cloud of virtual photons surrounding the electron (and the positron), both its charge and its mass change their numerical values. Since electrons and positrons interact with the radiation field at all times and in all circumstances, it is only these altered values of the mass and charge that are ever observable, and every experimental prediction must be expressed in terms of them rather than in terms of the unobservable *ur*-values that they would have in

---

[14] Since photons have no rest mass, there is no minimum energy needed to create them, as there is for pair creation, and unlimited numbers of softer and softer photons may always be emitted. In contrast to other infinities which originate from high energies, i.e., short wavelengths, and which are called "ultraviolet" divergencies, this effect leads to "infrared" divergencies, which, along with the former, have to be eliminated by renormalization and other devices.

the absence of coupling to the radiation field: this is called *renormalization*. It happens that QED leads to a number of divergent integrals, all of which are unambiguously removed by the process of renormalization, but the replacement of the unobservable "unrenormalized" mass and charge by the observable ones would be necessary even if this were not the case.[15] For the photon, the continual creation and re-annihilation of oppositely charged pairs leads to a "polarization of the vacuum," which would give the quantum of radiation a non-zero renormalized mass but for the enforced requirement of gauge invariance.

Left over after the renormalization procedure are the observable predictions of QED, which are experimentally confirmed to an astonishing degree of accuracy: the probability of the Compton effect, the anomalous magnetic moment (slightly different from a Bohr magneton) of the electron, the Lamb shift in atomic energy levels, the probability of scattering of light by light (being extremely small, this probability has been experimentally confirmed only rather indirectly), and many other phenomena.

FEYNMAN DIAGRAMS

All the predictions of QED are based on calculations employing perturbation theory, that is, series expansions in powers of the fine structure constant α. Since this constant has the small numerical value of approximately $1/137$, the first one or two terms in such series usually suffice. (A good thing, too. Since even the leading terms are usually complicated integrals, the higher-order terms require usually heroic efforts to calculate.) At the same time, each of the expressions in the perturbation series may be associated uniquely with a *Feynman diagram*, which appears to give it a highly intuitive interpretation in the language of particles—the

---

[15] Unfortunately, it turns out that assuming the unrenormalized mass to be zero, so that the observable mass would be entirely generated by interaction with the radiation field, does not work, just as classically the mass of the electron cannot be assumed to be completely of electromagnetic origin.

97

virtual creation and annihilation of electron-positron pairs or of photons, and of any number of combinations of these. For some physicists, this particle language is, indeed, fundamental, as it was for Feynman and Heisenberg. It should be clear from my presentation that I am advocating the opposite point of view: the quantum field is fundamental, as it was for Schwinger, and the appearance of "particles" is one of its secondary manifestations. The electrons of QED, of course, interact with one another via the electromagnetic field; electrons repel and scatter one another, attract positrons, and even form positronium with them. However, to think of this interaction as caused by the virtual emission and reabsorption of photons is at best a metaphor and should not be considered a serious physical intuition. Similarly, because of the nonlinearity of the QED equations coupling electrons to the electromagnetic field, photons interact with photons, producing the Delbrück scattering of light by light mentioned earlier. The Feynman diagram for that is shown in Figure 4.1, but again, this particle interpretation must not be taken too seriously. It should be remembered that all such effects would exist and be even stronger if the numerical value of α were much larger, so that perturbation theory would be powerless, as is the case with all quantum field theories other than QED. If we were smart enough to solve the nonlinear equations without resorting to series expansions, no one would use Feynman diagrams in field theory. Intuitively appealing as these pictures may be, they would not be useful for calculations and would be employed only by physicists who prefer thinking of reality in terms of particles rather than fields; the two points of view would have little in common.

OTHER PARTICLES

The spectra of the Hamiltonians of quantum fields give rise to a great variety of discrete entities, some of which may be viewed as bound states of the basic quanta, others more conveniently as new objects. Positronium falls into the first category. This particle may be regarded simply as a bound state of an electron and a

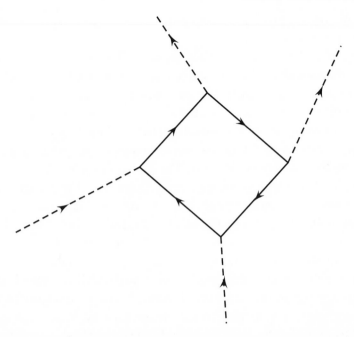

Figure 4.1. The simplest Feynman diagram for the scattering of light by light. The dashed lines represent photons, the solid lines electrons and positrons.

positron, but if the "radiative corrections" of its energy levels are taken into account, which means that the ubiquitous cloud of virtual photons—notice how we cannot escape the language of Feynman diagrams—and their concomitant vacuum polarizations, etc., are included in the picture, it becomes clear that such a view is no more than an approximation, albeit a close one. In the field theory of the strong interactions, quantum chromodynamics, where the basic quanta are quarks and gluons—analogous to electrons and photons in QED—"bound states of three quarks" and "bound states of a quark and an anti-quark" appear, the former constituting baryons such as neutrons and protons and the latter, mesons. Since the interaction in this case is not weak, these approximations are not nearly as valid as they are for positronium, but they are useful, nonetheless. The theory also predicts "glueballs," viewable as bound states of gluons, for which there is no analogue in QED because the effective interaction between photons is too weak to

99

be able to bind them. But these particles have not been experimentally observed yet.

The most basic theory of elementary particles we have at this time, the "standard model," encompasses both QCD and, with the electroweak theory, QED. Thus the standard model leads both to positronium and to the nucleons as "bound states" of basic quanta, and the variety of virtual processes that are being ignored in such a simple picture now becomes even greater. Other particles generated by the standard model are, of course, all the nuclei and even atoms and molecules. As we acend this hierarchy, it becomes a better and better approximation to regard them as bound states— nuclei as bound states of nucleons, atoms as bound states of nuclei and electrons, and molecules as bound states of atoms—but approximations they remain nevertheless. On the other hand, the standard model also generates a new particle, the Higgs boson, which appears in the spectrum as a result of a spontaneously broken symmetry and cannot be viewed as a bound state of others.

Some of the "particles" produced by field theory, the neutron being only one of them, are unstable. Such entities, of course, do not, like the stable states, appear as discrete points in the spectrum of the Hamiltonian, but they are visible as *resonances* in the S matrix. In other words, they manifest themselves as more or less sharp peaks in scattering and reaction cross sections of other particles, because the possibility of forming a long-lived state causes both a delay in the emergence and a deviation in the path of the interacting products. Usually these resonance peaks can—in theory—be made infinitely sharp, i.e., of zero width, by letting some interaction vanish.[16] In that case there will then be a stable particle in the spectrum, but the presence of the interaction opens up an escape route, allowing the otherwise stable entity to decay. Conversely, in a collision of the decay products, the long-lived entity that makes up the unstable particle of half-life T is formed temporarily, leading to a sharp peak of width $\hbar/T$ in the collision cross section.

---

[16]This is, however, not always so. Some sharp resonances do not owe their sharpness to the presence of some weak interaction, the vanishing of which would make the state stable. For a theoretical example of this, see pp. 548ff of my book, *Scattering Theory of Waves and Particles* (New York: Springer-Verlag, 1982).

The excited states of atoms are simple examples. In the absence of the electron's coupling to the radiation field these states would be stable, but in the presence of this coupling the excited levels turn unstable and the atom decays to the ground state by emitting a photon. At the same time, the scattering of electrons by ions exhibits sharp resonance peaks at the energies of excited levels, whose widths equal $\hbar$ divided by the level's radiative lifetime.[17]

CONCLUSION

Consider what has happened to the field concept. Originally, particles were the primary objects and the classical field was introduced as a condition of space to account for their interactions. The "quantization" of this classical field then gave rise to the quantum theory of fields, which, taken on its own terms, requires no a priori assumption of particles. In contrast to the classical field, the quantum field is autonomous as a condition of space everywhere, and quanta or "particles" are among its observable manifestations. The principal properties of these quanta are energy, momentum, and angular momentum (spin). What is more, each of them satisfies the canonical relativistic energy-momentum relation of a particle, defining its mass. However, these quanta lack many of the attributes of classical particles: indistinguishable from others of the same kind, they have no identity; they have no "substance" or size in the usual sense; and they have locations and trajectories, in any meaningful sense, only under very special circumstances. As a result of their emergence from the quantum field, they obey the rules of ordinary quantum mechanics, so long as their energies are low enough to prevent the creation of pairs and the emission of radiation is neglected. They exert forces on each other that can be approximately described by local potentials, which can be calculated from the underlying field theory. These are the objects—if that's what they may be called—which we think of as making up

---

[17] In actual practice, the experimental width will be much larger because of additional effects such as thermal motion; my example is meant schematically.

our observable world, the nuclei of atoms, the molecules forming solids, liquids, and gases, the chair we sit on, the house we live in, the stars we observe in the sky. The question of "reality" that inevitably arises from this will be the subject of discussion in a later chapter. But first we should look at the important concept of symmetry and its use in physics.

APPENDIX ON HARMONIC OSCILLATORS

The Hamiltonian of a one-dimensional simple harmonic oscillator of frequency ω is given by[18]

$$H = \frac{1}{2}[p^2 + \omega^2 q^2],$$

and its eigenstates and energy spectrum can be obtained by an elegant operator method (originally due to Dirac) as follows. Define the operator

$$a = \sqrt{\frac{\omega}{2\hbar}}\left[q + \frac{ip}{\omega}\right],$$

whose commutation relation with its hermitian conjugate is easily found from that between p and q, $[q, p] = i\hbar$ to be

$$[a, a^\dagger] = 1. \tag{1}$$

Also define the hermitian operator $N = a^\dagger a$, so that the Hamiltonian can be expressed in the form

$$H = \hbar\omega(N + 1/2). \tag{2}$$

At this point we may forget about q and p and pay attention to equations (1) and (2) only.

From equation (1) it is easy to obtain the commutation relations

$$[N, a] = -a, \quad [N, a^\dagger] = a^\dagger. \tag{3}$$

---

[18] Here q and p are the oscillator's coordinate and momentum.

Suppose, then, that $|n\rangle$ is an eigenstate of N with the eigenvalue n, $N|n\rangle = n|n\rangle$. It now follows from (3) that

$$Na^\dagger|n\rangle = (n+1)a^\dagger|n\rangle \tag{4}$$

and

$$Na|n\rangle = (n-1)a|n\rangle. \tag{5}$$

These equations imply that unless $a|n\rangle = 0$, $a|n\rangle$ must be an eigenstate of N with the eigenvalue $n-1$, and unless $a^\dagger|n\rangle = 0$, so must be $a^\dagger|n\rangle$ with the eigenvalue $n+1$. Thus if n is an eigenvalue of N, so are $n-1, n-2, \ldots$ and also $n+1, n+2, \ldots$. However, N is a positive semidefinite operator, $(\psi, N\psi) = (\psi, a^\dagger a\psi) = (a\psi, a\psi) \geq 0$; therefore, the downward steps in the spectrum must terminate before becoming negative, and the only way in which this can happen is if for the smallest eigenvalue n_0 we have $a|n_0\rangle = 0$. But since $N = a^\dagger a$, this implies $N|n_0\rangle = 0$ and therefore, $n_0 = 0$. The upward steps, on the other hand, cannot terminate because $\|a^\dagger|n\rangle\| = 0$ would conflict with (1).

We therefore conclude that the eigenvalues of N are all the nonnegative integers: $n = 0, 1, 2, \ldots$, which means the eigenvalues of H are

$$E_n = \hbar\omega(n + 1/2), \tag{6}$$

and the ground state, with the "zero-point energy" $\hbar\omega/2$, is determined by the equation $a|0\rangle = 0$. The nth normalized eigenstate $|n\rangle$ must be a multiple of $(a^\dagger)^n|0\rangle$, since $a^\dagger|n\rangle = c|n+1\rangle$. But the last equation implies by (4) and (1) that

$$|c|^2\langle n+1 \,|\, n+1\rangle = |c|^2 = \langle n|aa^\dagger|n\rangle = \langle n|a^\dagger a + 1|n\rangle$$

$$= \langle n|N+1|n\rangle = (n+1).$$

Therefore, we may choose $c = \sqrt{n+1}$, i.e., $a^\dagger|n\rangle = \sqrt{n+1}|n+1\rangle$, and, by repetition,

$$|n\rangle = \frac{(a^\dagger)^n}{\sqrt{n!}}|0\rangle.$$

The result (6), that the eigenvalues of H are equally spaced upward from a ground state of energy $\hbar\omega/2$, holds for the quantized form of any sinusoidal function $\sin \omega t$, since the equation $\ddot{a} = -\omega^2 a$ satisfied by this function can always be derived from a Hamiltonian of the form (2) by

$$\frac{\partial a}{\partial t} = \frac{i}{\hbar}[H, a]. \tag{7}$$

The operators $a^\dagger$ and a are called *raising* and *lowering operators*. Because of the equal spacing of the energy levels, the nth state lends itself to the interpretation that it is a state of n *quanta*, each of which has the energy $\hbar\omega$. With this interpretation in mind, a and $a^\dagger$ are then called *annihilation* and *creation operators* of quanta, respectively, $|0\rangle$ is the *vacuum state*, and N is the *number operator*, which counts the number of quanta.

In the quantization of the electromagnetic and other integral-spin fields, after spatial Fourier transformation, each Fourier component of the field turns out to satisfy a commutation relation like (1). On the other hand, the quantization of a field of half-integral spin requires the Fourier components to satisfy *anticommutation relations* like

$$\{a, a^\dagger\} = 1, \quad \{a, a\} = 0, \tag{8}$$

where $\{a, b\} = ab + ba$. (We still have the typical harmonic-oscillator equation $\ddot{a} = -\omega^2 a$ from (7) if the Hamiltonian is of the form (2), with N again defined as $N = a^\dagger a$.)

The second equation in (8) implies that $a^2 = 0$; therefore, we find from (8) that $N(N - 1) = 0$, which implies that N can have only the two eigenvalues 0 and 1. For the eigenstate $|0\rangle$ we have $\|a|0\rangle\|^2 = \langle 0|a^\dagger a|0\rangle = \langle 0|N|0\rangle = 0$, and hence

$$a|0\rangle = 0,$$

just as in the previous case. To get to $|1\rangle$, the operator $a^\dagger$ has to be applied; by use of (8) we find $Na^\dagger|0\rangle = a^\dagger|0\rangle$, and if $|0\rangle$ is normalized to 1, also $\|a^\dagger|0\rangle\|^2 = \langle 0|aa^\dagger|0\rangle = \langle 0|1 - a^\dagger a|0\rangle = 1$, so we can conclude that

$$a^\dagger|0\rangle = |1\rangle.$$

Furthermore, we easily find that $a^\dagger|1\rangle = 0$ and $a|1\rangle = |0\rangle$. In other words, the operators a and $a^\dagger$ again act as lowering and raising operators, or as destruction and creation operators of quanta. However, there are only two states, the vacuum state and the one-particle state. Instead of creating a second particle in the state $|1\rangle$, the operator $a^\dagger$ annihilates it: the particles of half-integral spin are fermions, obeying the Pauli exclusion principle.

Symmetry in Physics

SYMMETRY HAS ALWAYS played a role in our attempts to understand nature, guided at first by religious or esthetic points of view, later by a purely scientific perception. It was the symmetry of divine perfection or celestial beauty that dictated the motions of the heavens. Since the circle was regarded as the most ideal geometrical figure, reflecting God's glory, it was manifestly the only orbital shape suitable for a planet. Galileo was persuaded by his esthetic distaste for the elongated figures of "mannerist" painters and sculptors to reject Kepler's laws, which replaced the circles with ellipses.[1] But after Newton, who still believed in a God-given absolute space, religious considerations would fade out of science, while the use of esthetic criteria underwent a decided change.

SYMMETRIES IN A NEW GUISE

Instead of looking for esthetic perfection and concomitant symmetries in the specific motions or shapes presented to us by nature, we seek them in the laws governing the orbits and configurations. Since the specific ways in which planets or other objects move are determined by differential equations and initial conditions, the first expressing the laws and the second the accidents of history, symmetries and other simplicities we might expect in nature are bound to express themselves in the equations, but not necessarily in their solutions. The perfect rotational invariance originally embodied in the shape of a circle instead becomes a property of the gravitational law, together with the absence of a preferred direction contained in the equations of motion. That a symmetry of the underlying equations does not necessarily lead to the same

[1] Gerald Holton, *Einstein, History, and Other Passions* (Reading, Mass.: Addison-Wesley, 1996), pp. 97ff.

symmetry in their solutions results in "ugly" ellipses—a "spontaneously broken symmetry," as we would call it now.

Of course, we don't have to ascribe the choice of rotationally symmetric laws of gravity and of motion to esthetic preferences, and I don't claim that such were Newton's motivations; the most plausible explanation of this choice is certainly *simplicity*. In the absence of a good reason to introduce a preferred direction, rotational invariance is surely the simplest choice, and simplicity is a powerful criterion for physical theories. The important point is that symmetries, for whatever reason they are introduced, now are expected to express themselves not in the world as we directly experience it but in the underlying laws and theories, and in these they have been playing an increasingly central role in physics.

THE CASE OF PARITY

What, then, do we mean by *symmetry*? As I have already implied above, a symmetry means the same as an *invariance* under a certain operation. The simplest example of a symmetry, and one that, for a long time, all fundamental theories had been expected to incorporate, is invariance under reflection. An inversion of the direction of an odd number of coordinate axes cannot be accomplished by a rotation, as the inversion of an even number of axes can. Thus, in three dimensions, a reflection on a plane, which changes the sign of one direction, perpendicular to the plane, is fundamentally different from a rotation: the mirror image of your right hand is a left hand, and they cannot be rotated into one another. Inversion through the origin in three dimensions, in which all three coordinate axes reverse their signs, changing a right-handed coordinate system into a left-handed one, differs from mirror reflection only by a rotation (in which two axes change sign).

In classical physics, perhaps the most prominent example of the relevance of *parity*, the behavior of a function under a change of sign of its argument, occurs in the standard classification of the elementary patterns of electromagnetic radiation: the difference between electric and magnetic multipole radiation of a given order (angular momentum) is their opposite parity. The Maxwell equations, in fact, have a peculiar asymmetry with respect to inversion, which expresses the experimental fact that, so far as we know, there are no monopoles in nature. If there were, we would be faced with a choice of either defining the electric charge or the monopole strength as invariant under inversion; the mirror image of one of them would have to have the opposite sign, and it would be a matter of definition which one. As it is, we do not have to make this decision, because in the absence of monopoles, it is obviously much simpler to define the electric charge to be invariant, and as a result, the electric field to be a polar vector, the magnetic field to be axial.

In quantum mechanics, the tool for inversion is an operator—the *parity operator P*—that inverts the signs of all cartesian components of the arguments of a function. If a function is invariant under a change of sign of all its variables, so that $Pf = f$, we say it has *parity* +1; if it changes sign under such a change, so that $Pf = -f$, its parity is −1. The equation $HP = PH$, or equivalently, $PHP = H$ (note that $P^{-1} = P$), thus expresses the invariance of the Hamiltonian H of a given system. But this same equation also implies that P is a constant of the motion: *reflection symmetry implies that parity is conserved,* or equivalently, *parity is a good quantum number.* Since $P^2 = 1$, the eigenvalues of the parity operator are necessarily ±1; so if a system whose time development is determined by a reflection-invariant Hamiltonian, $H = PHP$, is in an eigenstate of parity at one time, it will remain in the same eigenstate. Note, furthermore, that unless a stationary state is degenerate, it has to be in an eigenstate of parity, because if Ψ is an eigenstate of H, so is $P\Psi$, and with the same eigenvalue. (If there is degeneracy, eigenstates of H may always be chosen to be eigenstates of parity.)

Conservation of parity in fundamental physics had been an article of faith among physicists for a very long time: the laws of nature surely would not express a preference for right-handed over left-handed coordinate systems, or vice versa. As a result, all outcomes of experiments were expected to look the same if seen in a mirror.[2] In particle physics, this assumption leads to very specific prohibitions of certain reactions, since particles can be assigned "intrinsic" parities, multiplicative quantum numbers whose conservation implies that their product for all participating particles have to be the same before and after any reaction. This presented a conundrum in 1956: experiments showed that there appeared to be two unstable particles, called τ and θ, that had essentially identical properties, including their mass, but different decay modes; one of them decayed into two pions, the other into three. Since the intrinsic parity of the pion is -1, it followed, by parity conservation, that the τ and the θ had opposite parities, which, in view of their other apparently identical properties, was very strange; the τ and θ seemed to form a "parity doublet." T. D. Lee and C. N. Yang solved the τ-θ *puzzle* by suggesting that here was a case in which parity was not conserved.[3] However, instead of simply putting this solution forward in an ad hoc fashion, which would surely not have been accepted—as it was, they had to overcome the almost universal scorn of the most prominent physicists of the day[4]— they connected this instance of "weak decay" to others, the most prominent of which are the nuclear beta decays. After thoroughly combing the literature and finding no actual evidence based on tests[5]—in contrast to an assumption everyone was making—that beta decay conserved parity, they proposed a specific experiment, namely, to look for a forward-backward asymmetry in the beta decay of a sample that was spin-polarized by an externally applied

[2] In microphysics, subject to the laws of quantum mechanics, this does not mean that mirror symmetry has to hold for each individual event, but only for the statistical distribution of large numbers of them.

[3] T. D. Lee and C. N. Yang, "Question of parity conservation in weak interactions," *Physical Review* **104** (1956), p. 254.

[4] Pauli wrote he could not believe "God was a weak left-hander," and Feynman bet Norman Ramsey that parity conservation would be upheld.

[5] Actually, there had been some evidence for nonconservation of parity in beta decay almost thirty years earlier, but it had been ignored: see R. T. Cox, C. G. McIlwraith, and B. Kurrelmeyer, "Apparent evidence of polarization in a beam of β-rays," *Proc. Natl. Acad. Sci. USA* **14** (1928), pp. 544–549.

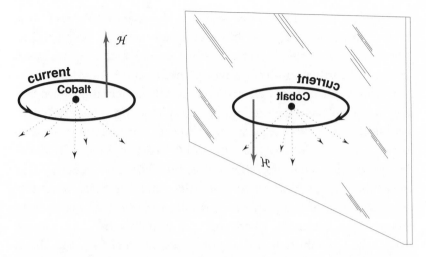

Figure 5.1. Schematic representation of the nuclear beta decay of Co⁶⁰ in the presence of a magnetic field produced by an electric-current coil, and its mirror image. The solid gray arrows indicate the right-handed screw sense associated with the current, and hence the direction of the magnetic field; the dashed arrows show the direction of the preponderance of the emitted electrons. The picture shows that the observed effect differs from its mirror image: parity is not conserved.

magnetic field. This experiment was very quickly carried out on Co⁶⁰ by Chieng-Shiung Wu and her collaborators,[6] and to the great surprise of almost everyone, the result came out positive: parity conservation was strongly violated (fig. 5.1). In short order, while all the skeptics wiped the egg off their faces,[7] Lee and Yang received the Nobel Prize, and (maximal) parity nonconservation was built into what became the standard theory of weak interactions, including beta decay.

---

[6] C.-S. Wu et al., "Experimental test of parity conservation in beta decay," *Physical Review* **105** (1957), p. 1413.

[7] In a letter to Emil Konopinski, dated December 16, 1956, Pauli concluded with the following "rule for reading of papers, which seems to me to hold in a very good approximation: what a modern theoretician says under the title 'universal,' consider to be simply nonsense" (private communication).

THE EXPRESSION AND ENFORCEMENT
OF OTHER SYMMETRIES

Let us now turn to more general spatial symmetries. By spherical symmetry of a law or an equation we mean no more than that it is invariant under the operation of rotating the physical system with respect to a given coordinate frame. At the same time, this implies that if the coordinate frame is rotated with respect to the system, the equations are form invariant: the laws have the same form, independently of the coordinate system in which they are expressed. *There is no preferred frame of reference.*

To enforce such a requirement, as you know, we classify all the dynamical variables and other quantities that enter into the physical laws according to their behavior under rotations of the coordinate system in which they are expressed—so-called passive transformations. Those that do not change at all are *scalars;* those with components that change under rotations like the cartesian coordinates are *vectors;* those with components that change like pairs of cartesian coordinates are *tensors of rank two;* those whose components change like $n$tuples of cartesian coordinates, *tensors of rank n.* Explicitly, if the cartesian coordinates of a point P with respect to the old system are x_i, $i = 1, 2, 3$, and those with respect to the new are

$$x'_i = \sum_{j=1,2,3} a_{ij} x_j,$$

where $\sum_k a_{ik} a_{jk} = \delta_{ij}$ (here δ_{ij} is the Kronecker symbol, equal to 1 if $i = j$ and zero otherwise), then the 3^n numbers $A_{i_1, \dots, i_n}$, given in the old system, are the components of a tensor of rank n if their values with respect to the new system are given by

$$A'_{i_1, \dots, i_n} = \sum_{j_1, \dots, j_n} a_{i_1 j_1} \cdots a_{i_n j_n} A_{j_1, \dots, j_n}.$$

If inversions are added to the rotations, we distinguish between scalars, which are invariant under reflections as well, and pseudo-scalars, which change sign, between vectors, whose components

111

change sign, and pseudo-vectors (or axial vectors), whose components do not, etc. Once all the quantities entering into the physical equations have been so classified, it is easy to enforce the requirement of form invariance: never add quantities that don't transform in the same way, and always be sure that the quantities on both sides of an equation transform equally under coordinate rotations and reflections.[8]

If a law, expressed in such a form-invariant manner, contains no externally fixed quantities and is in that sense *autonomous*, it is automatically rotationally symmetric. If there is an externally fixed quantity—a quantity not subject to the equations of motion—that is not a scalar, i.e., not invariant under rotations, the equations may still be form-invariant, but they are not necessarily invariant under rotations of the dynamical system with respect to the fixed object, a so-called active tranformation. For example, the equations of motion of planets around an ellipsoidal star are form invariant under coordinate rotations if the force components are appropriately expressed, but they are not invariant under an active rotation of the planetary system (the star being held fixed). When we say that the equations governing the entire universe are invariant under rotations, we mean that they are autonomous, with only scalar external quantities—quantities that are not subject to these equations—contained in them. The physical implication of rotational invariance of a system is that the results of all experiments done on it are independent of the orientation of the laboratory in which they are performed.

Another important symmetry is *translational invariance* which, again, holds if the dynamical equations are autonomous, containing no externally fixed quantities, or if the fixed parameters in the equations are themselves invariant with respect to translations in space. An example would be the Maxwell equations in vacuum, which contain the speed of light as a parameter. In a medium with a variable index of refraction, however, the equations are not

---

[8] Alternatively, there is the coordinate-free way of writing such equations that makes use of the exterior differential calculus, as is now customary in differential geometry.

autonomous and no longer translationally invariant. The same applies to invariance under time translation: unless the equations contain explicitly time-dependent parameters, the equations are invariant under translations in time. Thus the results of experiments on autonomous systems depend neither on the location of the laboratory nor on the time at which they are performed. These three symmetries or invariances—with respect to rotations, spatial translations, and time translations—embodied in the laws of physics are also sometimes expressed by saying that *the universe is isotropic and uniform, both spatially and temporally.*

Since Einstein's introduction of the special theory of relativity, an additional prerequisite has been added for any theory to be considered acceptable: it must be form invariant under Lorentz transformations. Since the latter are formally quite analogous to spatial rotations, this symmetry translates itself into a requirement similar to rotational invariance. The first step toward enforcing it is to classify all quantities according to their transformation properties under such transformations. However, spatial coordinates and time transform together, so that if the location and time of an event as seen in one laboratory are given by x^{μ}, $\mu = 0, 1, 2, 3$, where $x^0 = ct$, the same event observed in another laboratory moving with respect to the first with a uniform velocity v has the four space-time coordinates[9]

$$x'^{\mu} = \sum_{\nu} a^{\mu}_{\nu} x^{\nu},$$

LORENTZ

which is such that

$$(x'^0)^2 - (x'^1)^2 - (x'^2)^2 - (x'^3)^2 = (x^0)^2 - (x^1)^2 - (x^2)^2 - (x^3)^2.$$

A *scalar*, therefore, is a quantity that is invariant under Lorentz transformations, and the 4^n quantities $A^{\mu_1, \dots, \mu_n}$, as measured in the first laboratory, are the components of a four-tensor of rank n if in the second they have the values

$$A'^{\mu_1, \dots, \mu_n} = \sum_{\nu_1, \dots, \nu_n} a^{\mu_1}_{\nu_1} \cdots a^{\mu_n}_{\nu_n} A^{\nu_1, \dots, \nu_n}.$$

[9] This transformation may combine both a "boost" and a spatial rotation.

To enforce the form invariance of an equation under Lorentz transformations, we then simply have to check that no quantities are added to one another which do not transform equally and that both sides transform in the same way.[10]

We have, so far, merely reviewed certain procedures to facilitate the statement of physical laws in such a way that their invariance—or, more precisely, their form invariance—under Lorentz transformations and translations in space and time becomes manifest. The more interesting question is: what are the physical consequences of these invariances? Noether's theorem, one of the most important propositions of mathematical physics, contains these consequences. It states that to every invariance of the equations of motion—in the widest sense, including field equations—of a system under a continuous transformation, there corresponds a quantity that is conserved during the motion. Applicable both to classical and quantum physics, it constitutes a remarkable and far-reaching connection between the symmetries possessed by a theory and the conservation laws it implies. In quantum physics, in fact, its coverage is even wider because, on much simpler grounds, as we have already seen in the case of parity, it holds not only for continuous but also for discrete transformations, which are much more important in quantum mechanics than in classical physics.

The quantities that are conserved by virtue of symmetries under the transformations of spatial translation, time translation, and rotation are the *total momentum, energy, and angular momentum.* (Actually, the form invariance of field equations under Lorentz transformations leads to the conservation of a tensor of rank three, the covariant generalization of the angular momentum, some of whose components are those of the ordinary angular momentum.)[11] In other words, in a theory whose laws explicitly vary with time, energy will not be conserved; if it contains externally fixed obstacles in space, momentum conservation is lost, and if it is not rota-

---

[10] These things are stated here simply as reminders; they are not meant to be a short course in relativistic tensor algebra or tensor calculus. For further details, see, for example, J. D. Jackson, *Classical Electrodynamics* (New York: John Wiley & Sons, 1975).

[11] For the case of electromagnetism, see, for example, Jackson, ibid., p. 606.

tionally symmetric, angular momentum fails to be conserved. All the desirable and beautiful symmetries of a theory thus have important observable physical effects, and conversely: all the conservation laws, arguably the most important predictions of physics, have their origin in underlying symmetries.

Before discussing other specific consequences in more detail, however, we shall have to turn to a description of the mathematical tools needed for the most efficient treatment of symmetries in physics.

GROUP THEORY FOR SYMMETRIES

The set of transformations, like rotations, for example, or the permutation of certain variables, that leave an equation invariant (and for which we define the product AB of the two transformations A and B as first performing B, followed by A) possess the important mathematical characteristic of making up a *group*.[12] This is because the set has the following specific properties: (1) it is *closed* under multiplication: whenever the equation is invariant under A and B, it is also invariant under AB, so that AB is also a member of the set; (2) if A is in the set, so is its inverse: the inverse is defined as "undoing" the transformation, and if the equation is invariant under one, it must also be under the other; finally (3), the *unit* operation is in the set: the *unity transformation* E consists simply of doing nothing. Usually, the multiplication rules of a group are contained in its multiplication table, which defines all its mathematical properties.

There are many physically relevant groups of finite order, i.e., with a finite number of elements; examples are permutations of N objects, or special, discrete rotations, with or without reflections or inversions, that leave a specific regular polyhedron invariant. Inversions form a particularly simple group that has only two members, the reflection P and the unity E, and is abelian, of course.

[12]Since the result of two operations may depend on the order in which they are executed, this product is not necessarily commutative: it may be that $AB \neq BA$. If for all members of the group $AB = BA$, the group is called *abelian*.

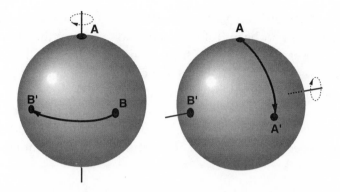

Rotation about axis through A   followed by   Rotation about axis through B'

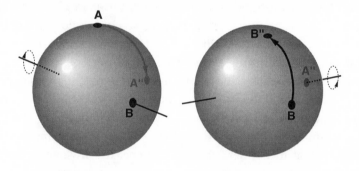

Rotation about axis through B   followed by   Rotation about axis through A"

Figure 5.2. Two rotations performed in different order are not necessarily the same. (Note that A″ is behind the ball and hidden by it.)

Other groups, such as translations and rotations that leave a certain plane tiling pattern invariant, are of infinite order. Then there are the Lie groups, whose elements depend, continuously and differentiably, on a number of parameters. Examples of these are unrestricted rotations and translations. Figure 5.2 shows that the group of rotations in three dimensions, called $O(3)$, which can be parametrized by the three Euler angles, for instance, is not abelian.

Some of the most important physical consequences of the existence of an invariance group in a theory flow from the fact that every element of a given group $\mathcal{G}$ can be represented by a finite-dimensional matrix, that is, there is a mapping between $\mathscr{G}$ and a group of $N \times N$ matrices, such that group multiplication in $\mathscr{G}$ is mapped into matrix multiplication and the matrices satisfy the same multiplication table as the members of $\mathscr{G}$. Of course, such an $N \times N$ representation is not unique because subjecting all the matrices M_i of a given representation $R(\mathscr{G})$ to the same similarity transformation XM_iX^{-1} results in another N-dimensional representation *equivalent* to the first. Now, if there exists a similarity transformation that puts all the matrices into the same block form, with lower-dimensional square blocks along the main diagonal and zeros everywhere else, the representation is called *reducible*. Because the individual blocks themselves then form representations, XRX^{-1} is nothing but the direct sum of the lower-dimensional representations formed by the blocks. A representation for which no such similarity transformation exists is called *irreducible*. These are the representations of greatest interest for any given group because all others can be formed out of them by taking direct sums—forming higher-dimensional matrices with the smaller ones on the main diagonal—followed by a similarity transformation.

The theory of group representations leads us to all the irreducible representations of a given group, or simply to the determination of their number and dimensionalities. In quantum mechanics, these dimensionalities, which group theory allows us to calculate relatively easily without having to use any properties of the Hamiltonian other than its symmetries, have enough physical significance to be of primary interest. Here is the reason.

As we have already seen in the case of reflections, in quantum mechanics, the group of transformations that leave a given physical system invariant can be represented by a group of operators that commute with the Hamiltonian. In fact, Eugene Wigner proved that any transformation leaving the equations of motion of a system invariant can be implemented by an operator which is either unitary or anti-unitary.[13] (The anti-unitary operators, which take i into $-i$, are needed only for time reversal.)

---

[13] E. P. Wigner, *Group Theory* (New York: Academic Press, 1959).

For example, if T, a member of an invariance group $\mathcal{G}$, is an operator that rotates all the particle coordinates of a system about a specified axis by a given angle, and the system is invariant under this rotation, then T must commute with H, so that $THT^{-1} = H$. This implies that if ψ is an eigenstate of H with the energy E, then $T\psi$ must be, too. But $T\psi$ is not necessarily a multiple of ψ— remember that the solution of the equation $H\psi = E\psi$ need not have the same symmetry as H. In fact, one can prove that if the eigenspace of H at the energy E is N-dimensional, i.e., the level is N-fold degenerate, $T\psi$ may be linearly independent of ψ, and T can be represented on that eigenspace by an N-dimensional matrix, as can all the matrices corresponding to the members of $\mathcal{G}$: they form a representation of that group. If this representation is reducible, the degeneracy is called *accidental* and may be removed or partially removed by a perturbation of H that has the same invariance group; if it is irreducible, the degeneracy is called *normal*: it cannot be reduced by any perturbation with the same symmetry as H.

Thus, finding the dimensionalities of all the irreducible representations of a symmetry group amounts to finding, without any further calculation, all the possible degrees of degeneracy that a Hamiltonian subject to that symmetry can normally be expected to have. This is the physically most important use of the theory of group representations in quantum mechanics. For instance, if, in the case of reflection invariance discussed above, that is the only symmetry of a given Hamiltonian, there is no normal degeneracy; in other words, in the absence of accidental degeneracy, all energy levels are non-degenerate, because this group is abelian and hence all the irreducible representations are one-dimensional.[14] In addition, the irreducible representations furnish information on what kinds of transitions from one level to another, caused by perturbations of a given symmetry, are *forbidden*: they provide the *selection rules*. There are many applications of this function of group theory,

[14] Since all the elements of an abelian group commute, they can be simultaneously diagonalized, which means the "blocks" obtained by the diagonalizing similarity transformation are all one-dimensional.

in crystallography for example, where the invariance group for a given crystal structure is usually of finite order. In a large number of important instances, however, the symmetry group is of the Lie type.

The rotation group in three dimensions, $O(3)$, is a case in point.[15] In this instance, we have a countably infinite number of irreducible representations, the lth, $l = 0, 1, 2, \ldots$, being of dimension $2l + 1$, thus leading to a $(2l + 1)$-fold normal degeneracy, the familiar m-degeneracy of rotationally invariant Hamiltonians.[16] The physical significance of the number l, of course, is the angular momentum. If we also admit "quasi-representations," in which a 2π rotation is represented by -1[17] rather than $+1$, and which are therefore double-valued, then the even-dimensional matrices come in as well, corresponding to angular momenta $s = \frac{1}{2}, \frac{3}{2}, \ldots$. The representations and quasi-representations of $O(3)$ can all be united by using the Caley-Klein parameters[18] to parametrize rotations in three dimensions, thus forming the "special unitary group" $SU(2)$, the group of (complex) unitary, unimodular 2×2 matrices, and their irreducible representations.[19] The rotations described by $SU(2)$, however, need not take place in ordinary, physical space but may refer to an abstract mathematical construct called "isospin space." Invariance of a given Hamiltonian under rotations in this "internal" space then leads to conservation of *isospin* and the degeneracy of isospin multiplets, such as the nucleon doublet, the neutron and proton. That the masses of the neutron and proton are not exactly equal is a consequence of the fact that the isospin symmetry is not exact; it is broken by the particles' coupling

---

[15] $O(3)$ stands for "orthogonal group in three dimensions." The lowest dimensional nontrivial, real representation of this group is formed by the 3×3 real orthogonal matrices, the familiar rotation matrices.

[16] On the other hand, if the Hamiltonian has only axial symmetry, there is no normal degeneracy because the rotation group in two dimensions (rotations about a fixed axis) is abelian.

[17] Here 1 is the unit matrix.

[18] See, for example, J. J. Sakurai, *Modern Quantum Mechanics* (Menlo Park, Calif.: Benjamin/Cummings, 1985).

[19] The Pauli spin matrices, together with the unit matrix, form a basis for a two-dimensional irreducible representation of $SU(2)$.

119

to the electromagnetic field: one is neutral, the other positively charged.

The case of a nonrelativistic particle in a Coulomb field is particularly interesting. As you may recall, the energy levels of a Hydrogen atom have not only the usual m-degeneracy, but also an l-degeneracy: the allowed energies depend only on the "total quantum number" $n = n' + l + 1$, where the "radial quantum number" n' can be $0, 1, 2, \ldots$. Is there an applicable symmetry group for which this is the "normal" degeneracy? *Yes.* Vladimir Fock proved[20] that the Coulomb Hamiltonian is invariant under rotations in a Euclidean four-dimensional space.[21] The irreducible representations of this group, $O(4)$, lead precisely to the multiplicities of the hydrogen energy levels as normal degeneracies.

GROUP THEORY IN PARTICLE PHYSICS AND FIELD THEORY

Just as the two nucleons form a doublet whose rest energy is (almost) degenerate, and this degeneracy is due to an $SU(2)$ symmetry in isospin space, so there are many other multiplets among the elementary particles, whose (almost) equal masses but otherwise identical properties are the result of symmetries in their own "internal" spaces. The groups $SU(3)$ and $SU(4)$ of unitary, unimodular matrices in three and four dimensions are particularly prominent in producing observed multiplicities of particles of (almost) equal masses.[22] Here, the actually observed mass splittings, however, are considerably larger—particularly in the case

[20] V. Fock, "Zur Theorie des Wasserstoffatoms," *Zeitschrift für Physik* **98** (1935), p. 145. See also M. Bander and C. Itzykson, "Group theory and the hydrogen atom," *Reviews of Modern Physics* **38** (1966), pp. 330 and 346.

[21] This is not to be confused with the Lorentz group relevant to the theory of relativity, which consists of rotations in four-dimensional Minkowski space; $O(4)$ leaves $x_1^2 + x_2^2 + x_3^2 + x_4^2$ invariant.

[22] For further information, see for example, D. B. Lichtenberg, *Unitary Symmetry and Elementary Particles* (New York: Academic Press, 1978).

of $SU(4)$—owing to much larger symmetry-breaking terms in the Hamiltonian. The theoretical prediction of the existence of the Ω^-, a negatively charged isospin singlet, was the crowning achievement of Murray Gell-Mann in his use of group representations to classify the many known baryons, and the experimental verification won him the Nobel Prize. This particle is pictured as consisting of three quarks belonging to three-dimensional representations of $SU(3)$, forming a singlet, analogous to two spin 1/2 particles (belonging to two-dimensional representations of $SU(2)$) in a singlet state.[23] In this picture, then, all the known baryons (without charm) are regarded as simply different states formed out of three $SU(3)$ triplets. (The addition of charmed particles requires enlarging $SU(3)$ to $SU(4)$.)

There is, however, a more pervasive symmetry that has come to play an important role in field theories: invariance under gauge transformations. All the Lagrangians used in quantum field theory contain the fields in at least bilinear form, that is, wherever a field ϕ appears, it is multiplied by its adjoint $\phi^\dagger$. This has the immediate consequence that the Lagrangian is invariant under a replacement of ϕ by $e^{-iq\theta}\phi$, so long as θ is independent of the space-time point x: there is *global* "gauge invariance of the first kind." However, if θ is taken to be a function of x, the terms containing $\partial_\mu\phi$ change into $e^{-iq\theta}[\partial_\mu - iq\partial_\mu\theta(x)]\phi$. These terms are not gauge invariant unless ϕ is coupled to a vector field $A_\mu(x)$ that appears together with all gradients in the combination $\partial_\mu + iqA_\mu(x)$, and whenever a gauge transformation of the first kind is performed, A_μ is changed by a "gauge transformation of the second kind" to $A_\mu + \partial_\mu\theta$. The existence of such a field, "minimally coupled" to ϕ, thus restores *local* gauge invariance. In order for the Lagrangian to remain invariant under local gauge transformations of the second kind, the "gauge field," when not coupled to ϕ, has to enter al-

---

[23] The Kronecker product of three $SU(3)$ triplets can be reduced to the direct sum of a decimet, two octets, and a singlet. Actually, Gell-Mann's prediction was originally based on reducing the Kronecker product of two eight-dimensional representations (the "eightfold way") into a direct sum of two representations of 8 dimensions, two of 10, one of 27, and a singlet.

ways in the combination $F_{\mu\nu} = \partial_\mu A_\nu - \partial_\nu A_\mu$, so that the extra term introduced in the gauge transformation of the second kind cancels out. The simplest form in which it can appear so as to be Lorentz invariant is as $F_{\mu\nu}F^{\mu\nu}$. This Lagrangian then leads directly to the Maxwell equations[24] for the electromagnetic field tensor $F_{\mu\nu}$. The quantity that, as a result of the gauge symmetry, is conserved, in accordance with Noether's theorem, is the electromagnetic current and, specifically, the charge q.

The line of reasoning used here shows that the assumption of *local gauge invariance* induces the theory to predict the existence of the electromagnetic field. This argument can be significantly generalized by no longer assuming that the group of gauge transformations is abelian. In other words, the field ϕ is assumed to have several components arranged as a column vector, and $\theta(x)$ is replaced by $\sum_\alpha T_\alpha \theta_\alpha(x)$, where each T_α is a square matrix that will act on ϕ. The transformations $\exp[-iq\sum_\alpha T_\alpha\theta_\alpha(x)]$ will now no longer commute for different values of $\theta_\alpha(x)$ and the group is non-abelian. The result is that the assumption of non-abelian gauge invariance thus generates new fields, whose quantization produces its own "gauge bosons."[25] All the modern field theories are now *gauge field theories*, and "supersymmetry" even connects bosons and fermions, or their fields, to one another.

In a certain sense, physics at the most fundamental level has returned to principles that governed it in the distant past: instead of searching for mechanisms that might determine the dynamics, as Maxwell did explicitly and others more implicitly, we begin by postulating large symmetries that underlie the physical world.[26] These symmetries or invariances—including, of course, Lorentz invariance—then generate, via the mathematics of group theory

[24] For details in the classical case, see Jackson, *Classical Electrodynamics*.
[25] For a good review, see Robert Mills, "Gauge fields," *American Journal of Physics* **47** (1989), p. 493.
[26] Not all the symmetries assumed are exact; some of them are, more or less weakly, broken. Furthermore, the most all-encompassing may hold only in the limit of enormously high energies or very small distances; at other scales they hold only approximately.

and analysis, the prediction of the existence of specific fields and the equations they must satisfy; finally, the fields, in their quantum form, give rise to the appearance of "particles." The ideal "theory of everything" might thus not be directly embodied in a set of equations, but in an all-encompassing *symmetry principle* from which the fields, the particles, and the dynamics would follow.

Causality and Probability

SCIENCE IS ALMOST unimaginable without the notion of causality. Yet what, exactly, is meant by causality has undergone considerable changes over the centuries. Of the four kinds of causes identified by Aristotle—material, final, formal, and efficient—we recognize only the last, and even that has altered its contours.[1] Whereas in Isaac Newton's time efficient causes appeared to bring about their effects by what amounted to a kind of physical effort, very few of us would go along with that today. Most physicists, consciously or unconsciously, have become followers of David Hume.[2] All we can say about the relation between cause and effect is that there is a *constant conjunction* between the two, and even that often gives way to a statistical correlation. (In quantum mechanics, the issue of constant versus statistical conjunction, of course, becomes particularly acute.) This, however, as we shall see, does not mean that causality plays a diminished role in physics.

What, then, do we mean by causality, and how can we tell the cause from its effect? In a certain sense, we feel that physical phenomena do not happen "on their own" so to speak, without any "reason" or cause. To be sure, the word "cause" is used in two different senses, one in which gravity "causes" the orbits of the planets and the falling of an apple, the other in which flipping a switch "causes" the light to go on. I prefer to use the word "explain" for the first sense and will focus on the second, in which causality has to do with events: two events may be claimed to be causally connected or not, and a theory can be said to be

---

[1] Formal causes, though we don't call them *causes*, still play a certain role in physics, for example, when we use symmetry considerations to explain why nature uses specific kinds of fundamental field theories and not others.

[2] It is perhaps well to remember that Immanuel Kant sought to rescue science from Hume's devastating critique of causality. However, I think we have survived that critique quite well without resorting to Kant's maneuver of making causality a mental necessity.

ALSO
SPACE
&
TIME?

124

causal if all repeatable events described by it have other repeatable events as causes. (Nonrepeatable events are of no interest to physics.)[3] It is precisely in this sense that quantum mechanics is regarded as *acausal*: we cannot assign a specific cause to the decay of a radioactive nucleus at a particular time. The clicking of a Geiger counter may be taken to be the quintessential appearance of a random sequence in nature, the clicks as unpredictable as the spin of the most honest roulette wheel. By *determinism*, on the other hand, which we have already discussed in chapter 2, I mean the idea that the present state of a physical system determines its later state, and it is important to distinguish this from causality. In my use of the word, quantum mechanics is as deterministic as classical mechanics; the difference between the two theories arises from their different definitions of *state,* (see chapter 2). In this chapter, however, I want to focus on the role of causality.

When we assert a causal connection between repeatable phenomena A and B, two questions arise: how do we define such an assertion, and how can we tell which is the cause and which the effect? Our first impulse is to say that B happens if and only if A happens. But then there is the possibility that B can be "brought about" by more than one cause; so it might be better to say that the occurrence of A is *sufficient* but not necessary for B. However, it is also possible that A can cause B or other effects, but B cannot happen without A; that would make A *necessary* but not sufficient. In the end, these are relatively minor complications because they can be eliminated by strict attention to the conditions (which may include *contributing* causes) in which the correlation between A and B is to be observed; the connection between the two may exist only in very specific circumstances that have to be isolated. When this has been done, we are back with the initial conclusion, somewhat sharpened, that, once the phenomena are sufficiently isolated from

---

[3] To the extent that physics is dealing with unique events, as in cosmogony, we are really talking about history rather than physics, albeit history that employs the tools of and is based on physics.

other influences, *A is both necessary and sufficient for the occurrence of B*, a relation that is symmetric between A and B.

For example, suppose you have set up your TV to have the signal come in through a VCR and the audio via the VCR and an amplifier. Suddenly you begin to hear a constant crackling noise when you watch a program. In order to find out which of the three components is the source of the trouble, you first switch off the VCR (which also disconnects the amplifier), so the TV is all on its own, and the noise has disappeared—the VCR (with amplifier) was necessary to produce the noise. Then you reconnect the VCR to the TV but turn off the amplifier, so that the sound comes directly from the TV set, and again you hear the crackling—a fault in the amplifier is not the cause and the presence of the VCR is sufficient to produce the disturbance; you have found the location of the cause: the presence of the VCR is both necessary and sufficient (other circumstances being equal) for the noise to appear.

There are many instances in which a causal link is established on a statistical rather than a universal basis: we find a strong statistical correlation between two events. Again, the phenomena need to be isolated as much as possible from other, extraneous influences, but it is important to recognize that here too, the causal relation is *symmetric* in the sense that the link by itself does not allow us to distinguish between cause and effect. We are all familiar with the ubiquitous question this symmetry raises when statistical correlations are found in social or medical contexts: is A causing B or is it the other way around? It should be clear, at the same time, that while the available evidence for a purported causal connection may, in some instances, be statistical only, what we ultimately mean when we say "A causes B"—once the circumstances are sufficiently isolated—is that the two are *universally* conjoined.[4]

---

[4] For probabilistic correlations, when dissipation enters into the picture, the symmetry between cause and effect may be broken. The mess on the floor may have been caused by many different kinds of accidents. That's why I want to distinguish causality from probabilistic relations.

WHICH IS THE CAUSE, WHICH THE EFFECT?

The second problem we face is how to tell the cause from the effect. The answer underlies every physics experiment: when I wiggle A, B is found to waggle. This is precisely where the experimental sciences like physics and chemistry have an enormous advantage over other sciences: we do not simply *observe* nature, we *experiment* on it. However, we have to be careful about what "wiggling A" means. The first answer that comes to mind is simply based on the idea that "I manipulate A the way I want to; I turn it on at will, and its strength or orientation is changed at my pleasure"; B, by contrast, happens without my direct intervention and its occurrence is correlated with A. In other words, the distinction between cause and effect is based on the assumption of free will. Free will, however, is a debatable notion; what if actions that I think are my own free choice are in reality determined by something else, which, for all I know, may include B?

A safer way of proceeding is to allow the occurrence of A to be determined by a random device, such as a roulette wheel, or by computer-generated random numbers. If there is a lingering suspicion that perhaps the random device is rigged or surreptitiously connected to B, we can arrange for long chains of independent games of chance to remove all such doubts and make certain that the occurrence of A is "under our control" in an extended sense. This is the crucial distinction between the two. If two phenomena are causally connected, *the cause is the one that is under our control.*[5]

Let us understand clearly that the distinction between cause and effect is not contradicted by the symmetry between events in time-reversal invariant physics. The motion of a ball flying from A to B might be interpreted as defining the event E_A at A as the cause

---

[5]If the universe is totally deterministic, so that any notion of control is illusory, no distinction between cause and effect is possible, and there is ultimately no such thing as experimentation. As physicists, we really have no choice but to assume, at least as a continuing working hypothesis, that the universe is not totally deterministic. Otherwise we would be out of a job, replaced by descriptive historians.

127

of the event E_B at B, thus defining E_B as the effect. But since the motion could equally well have gone from B to A, there seems to be no reason not to call E_B the cause and E_A the effect, rather than the other way around. This, however, is not how the words *cause* and *effect* are, in fact, used. We do not call the position and velocity of the moon today the cause of its position and velocity tomorrow. If you raise your arm and *throw* a ball at a puppet in a carnival, we call the throw the cause of the puppet's fall—a small error in your aim would have kept the puppet standing—not the puppet's fall the cause of your raised arm, even though the ball could have moved the other way. Reversibility of the signal or mechanism leading from cause to effect is irrelevant to the distinction between them.[6]

THE EFFECT COMES AFTER THE CAUSE

The next question that arises is the temporal order of cause and effect. You will have noticed that I did not define the cause as the earlier of the cause-effect pair, which might have been tempting. Unless the assumption of control is a complete illusion (in which case the distinction between cause and effect, other than their temporal order, is also an illusion), it is logically possible for an effect to precede its cause, i.e., for the future to influence the past; precognition is not *logically* impossible. In fact, however, this has never been reliably observed, and we have an enormous amount of accumulated evidence that effects *never* precede their causes. In other words, the universal temporal order of cause and effect is not a logical necessity but an observed, experimental fact of nature. (This statement is not meant, in any way, to put the universality of that temporal order in doubt, but simply to pin down its origin.)

---

[6] I therefore disagree with Stephen Hawking's argument in J. J. Halliwell et al., *Physical Origins of Time Asymmetry* (Cambridge, U.K.: Cambridge University Press, 1994), p. 346, where he writes: "So if state A evolved into state B, one could say A caused B. But one could equally well look at it in the other direction of time, and say that B caused A. So causality does not define a direction of time."

The temporal order of cause and effect, or what we may call the *causal arrow of time*, plays an important role in physics in a variety of contexts, and we shall return to it again in the next chapter. Perhaps the most ubiquitous of these roles is our preference for using *initial* conditions over final conditions in solving differential equations of time, as well as in other circumstances. When looking for the solution of an equation subject to given initial conditions, we are determining the development of a system under our control and causally influenced by us, which cannot be achieved by a final condition. Asking for the development of the system that led up to a given condition at a later time, though sometimes relevant— for some purposes in quantum scattering theory, for example—is generally of little physical interest because this development is not controlled or causally influenced by the condition. (Which is why the solution of the Schrödinger equation denoted by $\psi^{(-)}$ in scattering theory, useful as it may be, is usually regarded as physically less relevant than $\psi^{(+)}$.) The role of the time ordering of causality is prominent in the context of the theory of relativity as well.

One of the best-known consequences of the special theory of relativity is its prohibition of signals that travel faster than light. This prohibition, however, is not the result of relativity alone, but of the combination of the Lorentz transformation together with our confidence that the temporal order of cause and effect is inviolable. The argument, as you may recall, is that if a superluminal signal could be sent from A to B, as seen in reference frame 1, there would exist another frame 2, moving with respect to frame 1 with a speed less than the speed of light c, in which this signal would be observed as traveling from B to A, that is, it would "arrive" at A *after* it "left" B. Such a reversal of the sender and receiver, as seen by two observers, would allow the temporal order of cause and effect to depend on the coordinate frame, and this is regarded as intolerable.

The manner in which our usual notion of causality would be violated by the existence of superluminal signals can be understood in more detail by the following example (see figure 6.1), in which an observer in a laboratory L_1, stationary with respect to us, employs faster-than-light signals to send a message into his own past.

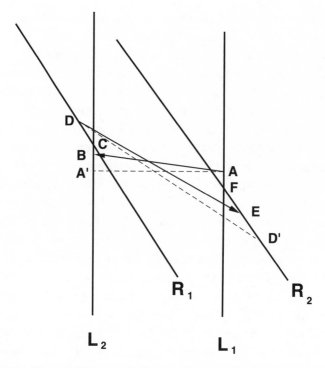

Figure 6.1. The world lines of two rocket ships and two stationary laboratories, communicating via superluminal signals. The dashed lines are the lines of simultaneity, as seen by the stationary observers and the rocket ships, respectively.

He begins by sending such a signal from the point A on his world line to another stationary laboratory L_2 at B. (The line of simultaneity for both of these stationary laboratories runs from A to A', so that they both agree B is later than A or A', and hence the signal was sent from A to B, and not the other way.) A bit later, as rocket ship R_1 passes, L_2 sends the message on to the rocket crew at C, who then pass it on by another superluminal signal to rocket ship R_2, which moves parallel to R_1 and at the same speed; the signal arrives at E. (The line of simultaneity for both rockets runs from D to D', and the two rocket crews agree that E is later than D and D'.) As R_2 passes L_1 at F, it sends the message back to the place where it originated, and it arrives *before* it was sent at A. The problem such a "causal cycle" produces can be highlighted by letting the message set off an explosion that destroys L_1, so that it cannot be sent at A, and the cycle cannot even start.[7] Following Einstein, we therefore conclude that superluminal signals must be impossible.[8]

Signals faster than light, then, are prohibited by a combination of special relativity and causality arguments.[9] This, of course, raises the question, what constitutes a *signal*? There certainly are physical phenomena that do travel faster than light; electromagnetic phase velocities always exceed c, for example. The connection with causality makes it quite clear what, in this context, is meant: a signal from A to B is anything that allows an effect at B to be the result of a cause at A. Schematically we may say that a signal from A to B is anything that allows a switch to be thrown at B on command from A. ("On command" here comes with the same caveats discussed above: either free will or a random device.)

As a result of the relativistic "no signals faster than light" prohibition, the temporal order of cause and effect leads to a more severe restriction than simply "no effect before its cause." If an event at A at the time t_1 causes another event at B at the time t_2, not only must t_2 be later than t_1 but it must be at least sufficiently much later for a light signal to be able to reach B from A. In other words, *the effect-event must lie inside the forward light cone of the cause-event*, or, put symmetrically, *no two events that lie outside one another's light cones can be causally connected.*

This prohibition plays a very special role in the relativistic quantum theory of fields. Let $\psi(x)$ be a quantum field as a func-

---

[7] For further discussion, see R. G. Newton, "Particles that travel faster than light?" *Science* **167** (March 20, 1970), pp. 1569–1574.

[8] It is sometimes claimed that the existence of a finite maximal signal velocity was one of Einstein's basic postulates for the derivation of the Lorentz transformation. This is historically incorrect.

[9] One form of signal could be made up of particles, but that these cannot approach the speed of light follows directly from the Lorentz transformation: unless they have zero mass, like photons, their energy would be infinite at the speed of light. Nevertheless, about thirty years ago there was some speculation that particles, dubbed "tachyons," which would always travel faster than light, might exist, and some experimenters searched for them. The idea was that quantum mechanics might prevent tachyons from being used as a signal; their existence might therefore not violate causality. None was ever found.

tion of the space-time point x. Any lack of commutativity of the field operator $\psi(x)$ (or of $\psi^\dagger(x)\psi(x)$) at the point x with $\psi(y)$ (or with $\psi^\dagger(y)\psi(y)$) at the point y can, by Heisenberg's indeterminacy principle, be interpreted as leading to the possibility of a disturbance of $\psi(y)$ (or of $\psi^\dagger(y)\psi(y)$) at y by a measurement of $\psi(x)$ (or of $\psi^\dagger(x)\psi(x)$).[10] Since such disturbances are restricted to travel no faster than light, one concludes that *the commutator* $[\psi(x), \psi(y)] = \psi(x)\psi(y) - \psi(y)\psi(x)$ *or the anticommutator* $\{\psi(x), \psi(y)\} = \psi(x)\psi(y) + \psi(y)\psi(x)$ *must vanish whenever x and y are outside one another's light cones.* (Similar conditions hold if one of the field operators is replaced by its hermitian conjugate.) Field theories that obey this so-called *causality condition* are generally designated *local* and are the only ones admitted.[11]

No Causality in Quantum Mechanics: Probabilities

Although, as I have argued earlier, quantum mechanics is as deterministic as classical physics, it undeniably has abandoned strict causality. Whereas determinism stipulates that the present state of an isolated physical system dictates its future state, causality refers to specific *events* causing other *events*; in that sense, classical physics was, in principle, governed by universal causality as

---

[10] For fermion fields, all observables are expressed in terms of $\psi^\dagger\psi$. That is why only this combination is relevant here.

[11] Since, as we shall discuss in more detail later, the interpretation of the indeterminacy principle as a measurement disturbance is seriously flawed, it is not really clear whether a lack of commutativity of measurable field quantities at different space-time points leads to the possibility of a signal. There are, however, other ways of arguing for the causality condition. One is based on the fact that unless the fields (or the mentioned products) on a spacelike surface commute, they cannot be independently specified, thus preventing the relativistic initial-value problem from being solvable. The other is that the causality condition leads to Lorentz invariance of the S matrix, though whether it is *necessary* for that Lorentz invariance is not clear. (Compare the statements on pp. 145 and 198 of Steven Weinberg's book, *The Quantum Theory of Fields*, vol. 1 [Cambridge, U.K.: Cambridge University Press, 1995].)

quantum physics is not. The future state of an unstable nucleus is determined if its present state is known—that is, determinism—but the time of its disintegration cannot be predicted on the basis of any prior measurement. It isn't possible to assign a cause to the event of disintegration; the best we can do is calculate its *probability* to occur in any given future time interval. As we discussed in chapter 2, the origin of the quantum-mechanical limitation to probabilistic predictions is inherent in the definition of a quantum *state*.

The important role played by probabilities, however, is not confined to quantum mechanics but exerts its influence throughout most of contemporary physics. Even classical dynamical systems, most of which behave chaotically, require probabilistic methods for their predictions. For many-particle systems like gases and liquids, this has been known, of course, for over one hundred and fifty years; that's, after all, where probabilities first entered physics. We therefore have to look a little more deeply into the meaning of probability.

THE MEANING OF PROBABILITY

The concept is as old as the existence of games of chance. It's a good guess that as soon as humans had leisure time, free from fighting for their necessities, they amused themselves with gambling.[12] Games of chance gave rise to the notion of probability, but the concept remained vague and ill defined, in part viewed as characterizing *opinion* rather than certain knowledge, and in part as an inherent charactistic of some kinds of events subject to chance.[13] The mathematical treatment of the subject of probability did not commence until the Renaissance with the work of Pascal

[12] The earliest pieces of archeological evidence for primitive forms of dice were found in Sumerian, Assyrian, and Egyptian sites; see Ian Hacking, *The Emergence of Probability* (Cambridge, U.K.: Cambridge University Press, 1975).

[13] This is the dichotomy between the epistemic and the aleatory notions of probability.

in the middle of the seventeenth century, brought to its peak of classical flowering by Laplace. The modern (objective) theory,[14] developed more or less in parallel with quantum mechanics, we owe primarily to Richard von Mises and Andrei Kolmogorov.

The break with the classical definition of probability (the ratio of "favorable outcomes" to "possible outcomes," based ultimately on an a priori postulated concept of "equally likely" occurrences) was made by von Mises in two papers published in 1919 in the *Mathematische Zeitschrift*, defining probability in statistical terms. Put simply, the probability for a specific event among a given (possibly infinite) number of other possibilities to occur is *defined* to be the limit of its relative frequency of occurrence in an infinite sequence of repetitions of the same "trial." The probability of throwing a three with a given die is not defined, as in the classical theory, to be the fraction this result represents among six equally likely possibilities—if the die is honest—but to be the relative number of times a three will occur if this particular die (honest or not) is cast an infinite number of times. The crux of the matter is that the repetitions serving as the basis for the definition of probability must form a *random sequence*—and there is the rub. While von Mises went to great lengths to define the concept of a random sequence, this notion, together with the question of the existence of a limit of relative frequencies in such an infinite sequence, lies at the heart of the mathematical objections that have been raised against his theory. These objections, however, were later largely overcome by the work of Per Martin-Löf,[15] and Kolmogorov's development of the

---

[14] The *subjectivist* theory of probability is based on notions such as "degree of belief," "our knowledge," or "available information." Its most complete version can be found in a large number of papers published by Bruno de Finetti. For a critical discussion of the subjectivist theory, see Karl Popper, *Realism and the Aim of Science*, in W. W. Bartley III, ed., *Postscript to the Logic and Scientific Discovery* (Totowa, N.J.: Rowan and Littlefield, 1983). There are a number of physicists who subscribe to it, giving rise to interpretations of quantum mechanics in which the observer plays a prominent role, but we shall ignore it in what follows. For a general overview of modern probability theory, see Jan von Plato, *Creating Modern Probability* (Cambridge, U.K.: Cambridge University Press, 1994).

[15] Per Martin-Löf, "The definition of random sequences," *Information and Control* **9** (1966), pp. 606–619, and "Complexity oscillations in infinite binary sequences," *Zeitschrift f. Wahrscheinlichkeitstheorie und verwandte Gebiete* **19** (1971), pp. 225–230.

modern mathematical theory of probability on the basis of, and almost coextensive with, measure theory assumes, for its application to the real world, von Mises's frequency definition.[16]

Von Mises's departure from the classical notion was motivated primarily by the use of probability in physics, where, up to that point, it had been limited almost entirely to statistical mechanics. Gibbs had introduced the idea of *ensembles* of systems of many degrees of freedom, such as gases, and had used them for the calculation of statistical averages, which then formed the basis of the thermodynamic laws. The low probability that the second law of thermodynamics would ever be found violated, for example, was calculated by means of ensembles of (macroscopically) identical systems, which may be thought of as nothing but simultaneous assemblages of the members of an infinite sequence. Equivalently, one may focus on the history of a given system and on the time sequence of the visits of the point representing it in its phase space to various (coarse) grains, and arrive at probabilities in that manner.[17] No matter, frequencies and ensembles were exactly the calculational basis for probabilities in physics, and von Mises simply made them their *definition* as well.

However, at the time of von Mises's introduction of the frequency definition, physics also went through the turmoil of the birth of the quantum theory, where probabilities were being used in novel ways. Like Brownian motion, the then recently discovered radioactive decay and photoelectric emission were not amenable to causal description, and von Mises was well aware of the importance of the probability concept in the new physics. But quantum mechanics, as developed by Werner Heisenberg, Erin Schrödinger, and Paul Dirac, was still waiting in the wings, and although the frequency theory was designed with physics in mind, its use in the ubiquitous application of probabilities in quantum mechanics was, and is to this day, considered objectionable by many physicists. The reason is not hard to find.

[16] See A. N. Kolmogorov, *Foundations of the Theory of Probability* (New York: Chelsea, 1950), p. 3.

[17] That these two ways of calculating probabilities lead to the same result is a consequence, or indeed the definition, of the ergodicity of the system.

135

Classical physics had always dealt with the description of individual systems; the only reason for introducing probabilities in the theory of gases was that we had no practical means of following the motion of each individual molecule, nor was it necessary to do so for the purpose of describing the macroscopic behavior of a gas. In quantum mechanics, on the other hand, probabilities were introduced at a basic level, and if they had to be interpreted via ensembles or frequencies, the state of a system would never seem to be a property of the individual system itself but, instead, a property of the ensemble to which it belonged. This bone is still sticking in the craws of many physicists. It is bad enough that the wave function of an electron does not describe the behavior of a specific electron itself, only of an infinite ensemble of them, but what about the wave function of the universe? Does it make sense, many have wondered, to talk about an ensemble of universes?

In addition, the frequency definition of probability, based as it is for most purposes on ensembles, is unpopular among some physicists because it suggests the question: what is quantum mechanics' analogue of the classical, Newtonian behavior of individual molecules that underlies the laws of statistical mechanics? Those laws arose from coarse graining[18] a phase space defined in terms of microstates that evolve causally; where are the microstates that would play an analogous role for quantum mechanics? An answer is provided by physicists like David Bohm and his followers, who introduce "hidden variables" precisely for that purpose. The ensemble view, shared by Einstein, is therefore associated in the minds of many physicists with an interpretation of quantum mechanics that most regard as unphysical. However, the two are not logically connected. Quantum-mechanical states are, indeed, in many respects analogous to macrostates, but there are, quite simply, *no microstates* in the quantum world. The quantum laws are probabilistic not because we lack information of *variables that exist but are beyond our knowledge;* they are *intrinsically* so. It is precisely

---

[18] For the definition of coarse graining, see chapter 7.

the phrase "variables that exist but are beyond our knowledge" that persuades us to call such interpretations unphysical.[19]

POPPER'S ATTEMPT TO AVOID FREQUENCIES

Karl Popper made a valiant attempt to avoid coming to the uncomfortable conclusion, drawn from the frequency interpretation of probability, that the quantum state of a system never refers to an individual system but only to an infinite ensemble by attaching the probability, defined though it may be through frequencies, to individual systems as a *propensity.* His reinterpretation was motivated not only by quantum mechanics but also by other uses of the probability concept for singular events. When speaking of the probability of a specific individual incident, having to confine oneself to those cases in which the event occurs as part of a sequence seems counterintuitive and unnecessarily restrictive.[20]

Of course, Popper's propensity to respond in a certain way could not be an invariable property of a system itself, because the system will act differently under different circumstances, depending upon the action or presence of others. The conditional probability $P(a \mid b)$ for a system to exhibit a under conditions b is the propensity of that system to behave that way under these conditions. The probability for a given neutron to decay in the next hour depends on whether it is moving freely or whether it finds itself inside a nucleus. But this simply makes propensity a *relational* property. The gravitational attraction that the sun exerts also depends upon the mass and distance of the body that it attracts. In

---

[19] To be fair, we should recognize, however, that many features of the Copenhagen interpretation of quantum mechanics may also be regarded as "unphysical," and the last word on what is the most appropriate version of the theory may not yet have been said.

[20] As I stated earlier, physicists are really interested only in repeatable events and not in truly singular occurrences, which are the domain of historians. An interpretation of the probability that "Caesar may have walked here" is of no great concern to us.

fact, Popper's idea was that propensity is a physical disposition of a system very much like a force, a tendency that "expresses itself in the relative frequency with which it succeeds in realizing the possibility in question."[21] Postulating propensity to be a physical property underlying the observed frequencies, he conjectured this property to be physically real in the same sense in which attractive or repulsive forces may be regarded as real. He even proposed to introduce *fields of propensities* analogous to force fields. In contrast to the frequency interpretation, which, in his understanding, made it necessary to regard any individual event to which a probability is assigned as a member of an actual sequence, he based his propensity on a *virtual* or *conceivable* sequence of events.[22]

Now it seems to me that while Popper's propensities are in part not very different from the notion of probability I am supporting, they include unnecessary metaphyical baggage, adding no new physical content to the probability concept. Many statements by von Mises notwithstanding, it is surely not necessary for the frequency theory to be interpreted as literally requiring the physical existence of an ensemble or a sequence in every case in which a probability is assigned. That was never required in statistical mechanics, either. Analogously, the physical meaning of an electric field is defined by means of its action on a test charge, but this does not mean that we cannot assert the existence of a field without actually placing test charges everywhere. We do not hesitate to assume an electric field even in remote places inaccessible to such testing.

Similarly, in order to give meaning to a probability, it is surely sufficient to have a virtual or conceivable sequence or ensemble in mind. That is why the mere impossibility of realizing an ensemble of universes does not appear to me to be a real obstacle in applying the frequency theory to probabilities for the universe as a whole, nor is it necessary to postulate "many (real) worlds" for such a purpose, as Everett's interpretation of quantum mechanics

---

[21] Popper, *Realism and the Aim of Science*, p. 286.
[22] Ibid., p. 287.

does.[23] The description of the state of a physical system by means of a state vector (or a wave function) in quantum mechanics, and the concomitant attachment of probabilities (all calculable from the state vector) to the state does exactly what Popper intends to accomplish with his propensities, except that the state vector itself is not a probability; instead it allows us to calculate the needed probabilities from it by well-defined rules. All of which persuades me that the frequency interpretation of probabilities is, at least for physics, still the most sensible there is, and there is no need for any reinterpretation.

STRANGENESS OF PROBABILISTIC THEORIES

Any physical theory that deals with probabilities rather than straight causal connections will necessarily exhibit certain strange features. It would be a good idea to look at some examples that have recognizable echos in quantum mechanics, but which should not be interpreted as implying that they exhaust the "strangeness" of quantum mechanics.

At a rather trivial level, take the following fanciful situation. Julie and Rachel, living apart in the same city, are watching the identical video program that is broadcast in black and white or in color, each with a 50 : 50 probability, and in English or Italian, also each with equal probability (the language and color are uncorrelated). The two women, independently, are allowed to turn on the sound or the picture, but not both. If Julie watches her screen and sees it in color, Rachel, also viewing her screen, will certainly see color too. But if Julie listens to the sound and hears Italian, Rachel will have only a 50 : 50 chance to see color. One might be tempted to ask, "How does Rachel's set 'know' whether Julie viewed or listened?" The correlation between the two events is characteristic

---

[23] H. Everett III, " 'Relative state' formulation of quantum mechanics," *Review of Modern Physics* **29** (1957), pp. 454–462.

of their probabilistic nature.[24] Here are some less trivial and more detailed examples of "strangeness," closer to actual physics.

Let's look at the notorious "collapse of the wave function," which superficially appears to be a strictly quantum-mechanical phenomenon but which has its analogue in any probabilistic theory.[25] The following will make this clear.

Suppose a system to have been in state a at the time t_1 and the prediction that it will be in state b at the later time t_2 is probabilistic, i.e., it has a probability p greater than zero and less than one. If a measurement on the system at t_2 ascertains whether it is in state b or not, the prediction of its further development will depend on the outcome of this measurement and will not be the same as if it had not taken place. That change is characteristic of the probabilistic prediction: the probability for the system to be in state c at the time t_3, later than t_2, given that it was in state a at t_1, differs from the probability for the system to be in state c at t_3, given that it was in state b at t_2, unless the probability for the system to be in state b at t_2 (given that it was in state a at t_1) was equal to 0 or 1. This effect is quite analogous to what is generally called the "collapse of the wave function" in quantum mechanics: the measurement or observation at t_2 suddenly alters the probabilistic prediction at t_3. Only in the case of $p = 1$ or $p = 0$ does the intermediate measurement have no consequence, nor does a "collapse" take place. The collapse represents the instantaneous change produced in the probability to find the system in state c at time t_3 from a probability conditioned by "given that it was in the state a at t_1" to one conditioned by "given that it was in the state b at t_2."

Here is another example of how counterintuitive probabilistic theories are in a way that is usually regarded as characteristic

[24] I am deliberately echoing here the EPR debate, which I already touched upon in chapter 2 and to which I shall return once more later. Einstein, Podolsky, and Rosen drew the conclusion from their hypothetical experiment that quantum mechanics could not be an exhaustive description of reality, which is quite correct in the case of our little TV fancy, because artificially neither Julie nor Rachel was allowed to both look and listen. The example is not meant to throw any light on that particular issue raised by EPR.

[25] Karl Popper, *Quantum Theory and the Schism in Physics* (Totowa, N.J.: Rowman and Littlefield, 1983), pp. 72–74.

of the quantum theory. Consider a classical system consisting of many particles of the same kind, treated by statistical mechanics, with its phase space—also called Γ-space—divided up by a given coarse graining into "stars."[26] Assuming a system at the time t_1 to have been in star G_1, the probability that at the time t_3 it will be in star G_3 is given by

$$P(G_3 \mid G_1) = \frac{\mu[G_3 \cap \varphi_{31}(G_1)]}{\mu[G_1]}, \qquad (1)$$

where $\mu[A]$ is the volume of the set A in the Γ-space and $\varphi_{31}(G_1)$ is the set to which the Hamiltonian flow takes G_1 during the time from t_1 to t_3.[27]

The assumption underlying formula (1) for the probability is that all initial microstates in a given macrostate are equally likely: the microstates correspond to points in the Γ-space, the macrostates to stars,[28] and the flow is assumed to start with a uniform distribution of microstates in the star corresponding to the initial macrostate. In reality, of course, a flow arriving in G_1 from some given earlier star and uniformly distributed there, will generally not arrive in G_1 filling it either completely or uniformly; it will have developed long, thin tendrils, some of which may snake through G_1 and fill a part of it sparsely (see fig. 6.2). The justification for the assumption of initial uniformity is either ergodicity, or, for probabilistic models to which such an argument does not apply, the absence of reasonable alternatives, unless we want to take the entire previous history of the system into account. This choice of a fresh start, so to speak, at any initial time, which turns the process into a Markov chain, has a number of important consequences.

---

[26] For the definition of stars, see chapter 7.

[27] The concept of Hamiltonian flow is analogous to that of a fluid flow, but in phase space. It establishes a mapping φ, defined by the Hamiltonian equations of motion, from any point X in phase space at the time t_1 to its image $\varphi(X)$ at a later time. So we may also think of $\varphi(G_1)$ as the image of G_1 under the mapping φ.

[28] In quantum statistical mechanics, the closest we can come to a representation of an analogue of microstates—which is not very close—is rays in Hilbert space.

G$_1$

G$_2$

Figure 6.2. The flow distorts the initial star G$_1$, so that at a later time, the image of G$_1$ has little overlap with any given star, G$_2$.

For one thing, the theory is obviously not invariant under time reversal, even if the flow is. To predict the future behavior of the system, i.e., where it will most likely be at some given later time $t_3 > t_1$, we must find the smallest Γ-space set G_3^0 made up of a union of stars that contains essentially all of $\varphi_{31}(G_1)$,[29] or such that the probability of finding the system in G_3^0 is very close to 1. On the other hand, the probability that a system starting out in the quasi mirror image $G_3^{0\prime}$ of G_3^0, obtained by reversing all the

---

[29] "Essentially all" means G$_3$ is such that $1 - \mu[\varphi_{31}(G_1) \cap G_3]/\mu[G_1] \ll 1$.

momenta while keeping the positions unchanged, would retrace its way back to G_1 under this flow reversal is $\mu[\varphi_{13}(G_3^{0'}) \cap G_1]/\mu[G_3^{0'}]$, which is very small. The flow arriving from G_1 fills G_3^0 sparsely and nonuniformly, but for the calculation of the reversed flow, $G_3^{0'}$ is assumed to be uniformly occupied.

In fact, the rules for probabilistic postdiction are different from those for prediction; they are not even uniquely defined. Given $t_1 < t_3$ and that the system at the time t_3 is found in G_3, are we looking for the star G_1 for which the probability $P(G_3 \mid G_1)$ for the system to end up in G_3 if it came from G_1 at t_1 is maximal? This question is fairly specific, but the maximum is likely to be quite broad and the postdiction almost worthless, more so if G_3 is not a star but a union of many stars. (On the other hand, if G_3 is a star, even the maximal $P(G_3 \mid G_1)$ is likely to be very small.) Or are we looking for the smallest union G_1^0 of stars such that essentially all of $\varphi_{31}(G_1^0)$ is inside G_3, i.e., such that the probability of finding the system in G_3 is very close to 1? The answer to this may be unique, but not very interesting.

Let us take a concrete example to illustrate the abstract argument. If a small gas-filled ampule is broken at the time t_1, releasing its contents into an empty surrounding container, at a (substantially)[30] later time t_3, its molecules are most likely going to be almost uniformly distributed in the larger vessel. On the other hand, if they start out being simply "almost uniformly distributed in the container" but with their momenta reversed, their chances of all ending up in the small region of the ampule are almost nil. The reason is that the phase-space set corresponding to "almost uniformly distributed in the container" (at a given temperature) has a much larger volume than that corresponding to the initial state of confinement to the ampule.[31] There is a very large number of microstates for which the molecules are "almost uniformly distributed" in the originally empty vessel, but a relatively small number of such states in which all the

---

[30] Meaning that $t_3 - t_1$ is larger than the relaxation time.
[31] Such discrepancies in size are particularly large for systems that are sensitive to initial conditions or have attractors.

molecules are near the small ampule and moving away from it.[32] As a result, the two probabilities are vastly different, and we have irreversibility.

However, we can also ask a postdictive question: from what small (ampule-sized) region, if any, did all the gas that ended up "almost uniformly distributed in the container" most likely come? The answer is not going to be very precise; there will be many such regions for which the probabilities of becoming "almost uniformly distributed in the container" are essentially equal. Or we can ask another postdictive question: what general earlier distribution of the gas molecules almost certainly led to their ending up "almost uniformly distributed in the container"? The answer now will not be very informative either: with probability very close to 1, the gas was "almost uniformly distributed in the container" (but slightly differently from now) at the earlier time too.

All of this, of course, raises the question of what the fundamental reason is for the asymmetric manner in which we treat predictions and postdictions in probabilistic systems. It seems clear to me that our ingrained expectations of causality are at the basis of the answer, together with our universal experience of causes preceding their effects. This is why we view the *initial* state to be under our control, populating the *initial* grain uniformly to arrive at probabilities for final states under the assumption of equal initial a priori probabilities. (Relying on ergodicity for this purpose is analogous to choosing a random device for making causal choices.) The course of the flow, the grains it visits and how densely it populates them, on the other hand, are the effects over which we have no direct control; therefore they occur later. To employ equation (1) for postdiction, with $t_1 > t_3$, would thus violate our sense of causality.[33]

---

[32] For a discussion of the reasons for this discrepancy, see the next chapter.

[33] From relativistic causality it is clear that predicted states have to be confined to the forward light cone of initial ones. If they are, there can be no disagreement among different observers as to which is the earlier one and whether we are predicting or postdicting. If the space projection of the star G_3 at the time t_3 in equation (1) lies entirely outside the forward light cone of the space projection of G_1 at the time t_1, $P(G_3 \mid G_1)$ must necessarily vanish; in fact, (1) makes sense only

The fact that the formula (1), prescribing the rule for prediction, cannot be used for postdiction, means that in classical statistical mechanics we have an analogue of the rule in quantum mechanics, that the state of a system, as ascertained by measurements, can be used for predictive but not for postdictive probabilities, which is often regarded as puzzling and counterintuitive.

Quantum mechanics is frequently criticized for another feature that is a general characteristic of probabilistic theories—the rules governing the process of "measurement" or observation are different from the laws obeyed by the ordinary flow $\varphi(G)$. At the moment a measurement is performed or an observation is made, the normal flow is in effect interrupted and a new state is prepared, again described by a uniform distribution in the coarse grain corresponding to the measurement outcome or observation. The counterintuitive results of this can be appreciated best by considering what happens when a system is observed at an intermediate time but the observation is ignored.

If at the time t_2 a measurement is performed on a given system that was in star G_1 at the time t_1, the probability of finding it in coarse grain G_2 and later, at the time t_3, in G_3 is $P(G_3 \mid G_2) \cdot P(G_2 \mid G_1)$. Suppose now that the outcome of the intermediate observation at t_2 is ignored; then the probability of finding the system in G_3 is

$$P'_{t_2}(G_3 \mid G_1) = \sum_{G_2} P(G_3 \mid G_2)P(G_2 \mid G_1) \neq P(G_3 \mid G_1),$$

where the sum runs over all stars G_2 that the system can possibly visit on the way from G_1 to G_3. (See fig. 6.3 for a simple example.) At first sight, this result seems to contradict the basic assumption underlying Bayesian probabilities, namely that $\sum_i P(A \mid B_i)P(B_i) = P(A)$ if the event A can come about only via

---

if all of the space projection of G_3 at t_3 lies inside the forward light cone of that of G_1 at t_1. However, we also have to remember that nonrelativistic theories, which are not explicitly constructed to be relativistically invariant, will not generally satisfy the strictures of Lorentz-invariant relativistic causality. In such theories, two observers in relatively moving frames may therefore disagree about what is prediction and what is postdiction.

145

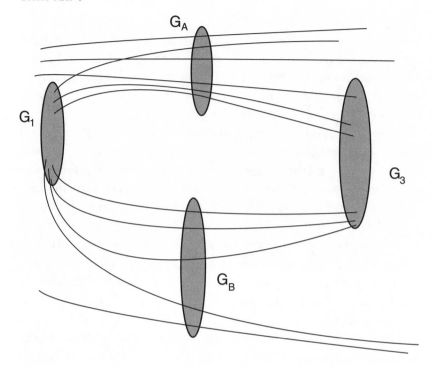

Figure 6.3. In this case $P(G_A \mid G_1) = \frac{3}{7}$, $P(G_B \mid G_1) = \frac{4}{7}$, $P(G_3 \mid G_A) = \frac{1}{2}$, $P(G_3 \mid G_B) = \frac{3}{5}$, which gives $P'(G_3 \mid G_1) = \frac{39}{70}$, whereas $P(G_3 \mid G_1) = \frac{5}{7}$.

the exhaustive set of mutually exclusive possibilities B_i. The reason for the discrepancy is that the stars G_2 can be reached not only from G_1, but from other parts of phase space, and the trajectories reaching G_2 from G_1 do not uniformly fill up G_2, as is assumed for calculating the probability $P(G_3 \mid G_2)$. It might be, for example, that the probability for the system to get from G_1 to G_3 via a particular grain G_2^0 is very small; nevertheless, there could still be a large contribution from G_2^0 to $P'_{t_2}(G_3 \mid G_1)$ because neither $P(G_3 \mid G_2^0)$ nor $P(G_2^0 \mid G_1)$ need be small.

Therefore, counterintuitive as this conclusion may be, the probabilistic prediction for a system that is unobserved between the times t_1 and t_3 generally differs from the prediction for the same system observed at an intermediate time, even if the outcome of this observation is ignored. The origin of the peculiarity of yielding different predictions for observed and unobserved systems is what might be called the "renormalization" of the initial state that necessarily occurs at the assignment of a conditional probability based on such an initial state.[34] It is here also that time-reversal invariance is lost and prediction differs from postdiction.

While classical statistical mechanics served as the model underlying our calculations, a similar difference between observed and unobserved systems—one of the notoriously counterintuitive results of quantum mechanics—may be expected to hold in other probabilistic models. In fact, it is sometimes asserted to be intolerable that, in quantum mechanics, the process of measurement or observation is not governed by the same laws as the time development of an unobserved system; something strange is claimed to occur during a measurement, which is responsible for the difference between observed and unobserved systems. However, the same is true for any system subject to probabilistic rules; this feature is not a special characteristic of quantum systems.

THE DECAY OF UNSTABLE STATES

An important special instance of the foregoing argument is that of the decay of unstable states. If the phase space is divided up into only two coarse grains of macrostates, those of the decayed and the undecayed system, the first of which is enormously more voluminous than the second, then the probability of finding the system undecayed at the time mt/n if observed at the time $(m-1)t/n$ shortly before (assuming $n \gg 1$) is given to an excellent approximation by $1 - t/(\tau n)$; and by a repetition of this argument, the probability of finding it undecayed at the time t, if observed at t/n and $2t/n \ldots$ and $(n-1)t/n$, is $(1 - t/\tau n)^n$. In the limit, as $n \to \infty$

---

[34] While any probabilistic theory must yield different predictions for observed and unobserved systems, I am not claiming that this "renormalization" is the sole source of this feature in quantum mechanics. It is, however, responsible for it in part.

147

this tends to the exponential decay law $e^{-t/\tau}$. Thus, any decaying probabilistic system subject to constantly repeated observation decays exponentially.

It is clear that the central characteristic of systems subject to this law is an *absence of memory*.[35] What is surprising, from this point of view, is that the very familiar argument presented (which, however, is usually offered without mentioning the assumption of frequent observation) leads to a result that agrees to an excellent approximation with what quantum mechanics predicts for an unobserved system, at least for certain lengths of time.

According to quantum mechanics—and in agreement with many experimental observations—most decaying systems follow the exponential decay law extremely well for many lifetimes,[36] but the decay law differs from an exponential for a short initial period and also for the tail end.[37] The details of the initial deviation depend to some extent on the method of excitation of the decaying state, but in most cases the decay starts out like $1 - (t/\tau_0)^2$, i.e., more slowly than linearly. This fact gives rise to what is known as the *quantum Zeno effect*:[38] if the system is subjected to repeated observation, and therefore to "renormalization" of the initial state, after very short time intervals—short compared to τ_0—it will always be found undecayed, like the proverbial watched pot that will not

[35] This is why the decay of an unstable elementary particle, if that concept makes sense, has to follow the exponential law exactly. No elementary system could have the structure needed for a memory.

[36] See, for example, chapter 19 of my book, *Scattering Theory of Waves and Particles*, 2nd ed. (New York: Springer-Verlag, 1982).

[37] However, some physicists have argued that, because of the constant interaction of decaying nuclei with perturbing surroundings, which act like effective observations, any derivation of an exponential decay law for unobserved systems that is only approximate instead of exact is practically irrelevant; it is the simple derivation for constantly observed systems that really matters. In view of the "quantum Zeno effect," this argument is of doubtful validity.

[38] Recall Zeno's paradox, which appeared to imply that the concept of motion was self-contradictory: any object that "occupies a volume of space equal to its own volume" is at rest; but a flying arrow occupies a volume equal to its own volume at every instance of its flight; therefore it must be at rest.

148

boil.[39] Contrary to the usual elementary derivations of the exponential decay law, therefore, which would be correct for a classical
probabilistic system, a quantum system under frequent observation will not decay exponentially; left unobserved, however, it will
follow that decay law accurately for long periods of time.

We now want to turn to a more detailed consideration of various
aspects of time's unidirectionality, a topic we have already touched
upon.

[39] B. Misra and E.C.G. Sudarshan, "The Zeno's paradox in quantum theory,"
Physical Review D **18** (1977), pp. 756–763; W. M. Itano et al., "Quantum Zeno effect,"
Physical Review A **41** (1990). p. 2295; "Comment" by L. E. Ballentine, ibid., **43** (1991),
p. 5165; and "Reply," ibid. **43** (1991), p. 5168; also B. Gaveau and L. S. Schulman,
"Limited quantum decay," *Journal Physics A: Math. Gen.* **28** (1995), p. 7359. For
a discussion of the difference between *continuous* and *intermittent* observations,
see L. S. Schulman, "Continuous and pulsed observations in the quantum Zeno
effect," *Physical Review A* **57** (1998), pp. 1509–1515.

Arrows of Time

ONE OF THE GREATEST puzzles of physics is the manifest discord between two facts: on one hand, all the fundamental equations and laws of physics are (essentially) invariant under time reversal;[1] on the other hand, we are all aware that at the macroscopic level, many physical processes flow in one time direction only. The unidirectional flow of time[2] is one of the most obvious features both of our consciousness and of the physical world. Here I want to examine this perhaps most prevalent "spontaneously broken symmetry" in nature, which near the end of the nineteenth century presented physics with one of its most profound challenges, and about which there are, to this day, strong disagreements among physicists.

THE FIVE ARROWS OF TIME

We have to draw distinctions between five different kinds of physical time arrows, and for each of them, if applicable, ask three separate questions: Why does the arrow exist for a given system? What accounts for the universal agreement of the direction in which this kind of arrow points for all systems? And finally, is there a connection between the direction of this arrow to those of the others? The five arrows of time are

1. The delay between cause and effect.

---

[1] The minuscule violation of time-reversal invariance any fundamental particle theory has to include in order to accommodate experimental results is of no real significance for our purposes. It is far too small to be related to the macroscopic unidirectional flow of time.

[2] To be distinguished from irreversibility, which we have already discussed in the previous chapter. Though not identical, the two are, of course, connected. In this chapter I shall focus primarily on the *arrow*.

2. The psychological awareness of the passage of time, pointing toward the future—some call this the biological arrow; I will call it *cognitive*.

3. The unidirectional flow of time embodied in the second law of thermodynamics.

4. The cosmological arrow defined by the expansion of the universe.

5. The direction of the time parameter used in physics, the directed fourth dimension of space-time, which, in contrast to the direction of a space axis, cannot be reversed by a physically realizable Lorentz transformation continuously connected to the identity (the orthochronous Lorentz group). We will see that the direction (which is, of course, conventional but has to remain consistent) of this time axis is also a player in the game because of the role of causality in conjunction with relativity, discussed in the last chapter.

The first arrow is defined by the time-ordering of cause and effect. That the temporal order of causes (as defined in chapter 6 in terms of *control* over their occurrence) and their effects seems to be universal, i.e., appears to be *always and everywhere the same*, has to be regarded as one of the most solidly established experimental facts of nature. Logically, there would, however, still be the possibility that this causal arrow points one way (relative to the direction of the physical time parameter, or fifth arrow) in one spatial region or period of history and in the opposite direction in another. If this were the case, for reasons explained in the last chapter, such discrepancies would have to be confined to space-time regions completely outside each other's (forward and backward) light cones. However, since the light cones of any two observers at space-time points A and B will necessarily intersect somewhere, implying A and B will lie inside the light cone of a point in the interior of the intersection, relativity theory guarantees that, physically, the time directions defined by the local causal arrows have to be in agreement in all of space-time.[3]

Regarding the occurrence of effects as *later* than their causes, on the other hand, is a definition based on our psychological sen-

---

[3] By agreement of the "time direction" I do not mean that the time axes have to be parallel, but that each of their positive time axes has to be inside the other's forward light cone.

sation of the direction of time, its *cognitive* arrow, pointing from "earlier" to "later," separating in our consciousness the past from the future. Two principal characteristics distinguish what will be from what has been: the future can be causally influenced, whereas the past cannot, and we have a memory or record of the past but not of the future. The first of these implies that the cognitive and causal arrows are directly connected: given that nature arranges the temporal order of cause and effect always to be the same, it is no accident that we regard effects as later than (in the future from) their causes, because what we regard as the past can no longer be causally influenced, whereas the events of the future can. The universality (both from person to person and in the course of time) of the direction of the cognitive arrow, then, rests on the experimental fact of the universality of the existence and direction of the causal arrow, in the following sense. It would be logically possible for the causal arrow not to exist at all, that is, for causes sometimes to occur earlier and sometimes later than their effects, or to exist, but to vary from system to system or from epoch to epoch. That in every epoch the causal arrow exists and does not vary from one system to another is a well-established experiential fact. In each time period, the causal arrow thus defines the psychological distinction between the past and the future. Moreover, since the direction of the causal arrow is universal in space and time, the distinction between past and future must have that property as well.

The second characteristic of the cognitive arrow—we have a record of the past but not the future—connects it to the third of our arrows of time, the thermodynamic (records are traces left by irreversible thermodynamic processes), whose direction, we shall see, is also finally determined by that of the causal arrow. (Moreover, that the traces of the past cannot be causally reached and altered is, again, due to the causal arrow.) Thus, both defining markers of the direction of the cognitive arrow of time are ultimately grounded in the causal arrow.

The time direction defined by the second law of thermodynamics, with which Boltzmann struggled a century ago, is still sometimes subject to misinterpretation. Here we have to come to grips with the strange phenomenon that not only irreversibility but a definite sense of a unidirectional flow of physical time seems to emerge from statistical mechanics, in spite of the fact that the underlying dynamics of the molecules is invariant under time reversal.[4] With his H-theorem, Boltzmann pointed the way to the solution of this problem, at least for ideal gases.

The H-theorem is proved[5] by means of Boltzmann's transport equation for molecules, based on the premise of "molecular chaos," his famous *Stoßzahlansatz*: *before* every molecular collision the momenta of the colliding molecules are uniformly distributed and independent of their positions; afterwards, of course, they become correlated. (A molecule is unlikely to be found in a region of space to which its momentum, after a collision, will not take it.) This assumption, which is sometimes portrayed as begging the question Boltzmann meant to answer, is quite analogous to the use of a uniform distribution in *initial* coarse grains for statistical (probabilistic) predictions, as we discussed in chapter 6, where I argued it is based, ultimately, on causality. Rather than begging the question, therefore, the *Stoßzahlansatz* merely connects the direction of the time arrow established by the H-theorem to that of causality. However, the meaning of the second law, or of the H-theorem, cannot really be understood without an examination of thermal fluctuations.

Coarse Graining

In order to remind ourselves of how the general increase of entropy of an isolated system of many degrees of freedom comes

[4] I will base my discussion on classical statistical mechanics, but the same problem arises, and is solved similarly, in quantum statistical mechanics. A good reference for a detailed discussion of the thermodynamic arrow of time is L. S. Schulman, *Time's Arrow and Quantum Measurement* (Cambridge, U.K.: Cambridge University Press, 1997).

[5] See, for example, Dr. Ter Haar, *Elements of Thermostatistics* (New York: Holt, Rinehart and Winston, 1966), or G. Wannier, *Statistical Physics* (New York: John Wiley & Sons, 1966).

about, let us begin by defining what is meant by *coarse grain-ing* a system's phase space, thereby at the same time defining its *macrostates*: to specify a system's macrostate is to specify the coarse grain containing the corresponding system point in its Γ-space.

Such coarse graining is based, to begin with, on equal-sized grains (to whose exact sizes, determined by the accuracy of the relevant measuring instruments, the final results are not very sensitive)[6] in the phase spaces of the individual particles. On the other hand, the coarse grains, or "stars," in the big phase space of the whole system, the Γ-space—the direct product of the six-dimensional phase spaces of the N constituent particles, and hence $6N$ dimensional—have extremely unequal volumes. Because we cannot macroscopically distinguish between different molecules of the same kind, a grain that contains, say, the point $\cdots \mathbf{p}_a, \mathbf{q}_a \cdots \mathbf{P}_b, \mathbf{Q}_b \cdots$ specifying the positions and momenta of particles a and b must also contain the point $\ldots \mathbf{P}_a, \mathbf{Q}_a \cdots \mathbf{p}_b, \mathbf{q}_b \cdots,$ which differs from the first by a simple exchange. As a result, for example, the volume of a star corresponding to all particles be-ing located in identical grains of their own phase space, with the remainder of the Γ-space empty, is minute compared to that of a star corresponding to all particles being in different grains of their own. (Since there are $N!$ ways of distributing N particles among N different grains, the ratio of the Γ-space volumes of these two stars for an N-particle system is $N!$.) The more widely the parti-cles are distributed over different individual grains, the larger is the corresponding star in Γ-space.

There are therefore enormous differences in volume between stars in Γ-space, corresponding to the vastly different "number of microstates" contained in them. (Note also that, because of the exchange invariance, individual stars may consist of disconnected pieces.) One macrostate of a gas at a given temperature in a given

---

[6] However, their space dimension should be small compared to the volume V of the container but large compared to V divided by the number N of particles, and their kinetic energy dimension should be small compared to the total energy E but large compared to E/N so that, if we represent each particle of the system by a point in the six-dimensional phase space, a coarse grain will, on average, contain many particles.

vessel would have all the molecules concentrated near one corner, with equal momenta, for instance, and another would have them all in different coarse grains of their own phase space, more or less uniformly distributed spatially and with a Maxwellian distribution of their momenta. The volume of the star corresponding to the latter is vastly larger than that of the star corresponding to the former.[7]

The Thermodynamic Arrow

In the course of time, the point representing a given system in its Γ-space will wander through all the stars on its energy hypersurface,[8] and, according to the ergodic theorem, the fraction of the time it will eventually have spent in star A is proportional to the volume $\mu(A)$.[9] The entropy S of the system at the time t may be defined as the logarithm of the probability of the macrostate in which it finds itself at the time t, which means that,

---

[7]There are other systems for which the sizes of coarse grains differ for different reasons. The grains defining the macrostate of a new artistic work of sculpture or a new car are relatively small, because minute nicks would be easily noticeable and we would reject them. On the other hand, the grains corresponding to the sculpture after being exposed to the elements for a millennium or the car after a collision, are enormously much larger, because we can make no distinction between one piece of weathered stone and another or between two piles of junk. This is what we mean by "order" versus "disorder." The undamaged piece of sculpture and the new car are in a *highly ordered* state: small changes would put them in different macrostates. The wreck and the ruin are in a state of disorder: we make no distinction between a wreck with a broken steering wheel and one with a broken axle among all the other damage, or between ruined torsos that, for all we can tell, might once have been a Venus or a Zeus. The star corresponding to a disordered macrostate is vastly larger than that of an orderly macrostate. (A pile of what appears like junk but in which small differences do matter may be a work of art made to look that way. It will be in a state of order and its star much smaller than that of real junk. Coarse graining depends on the focus.)

[8]If the system has other conserved quantities, the point representing it will remain on a correspondingly lower-dimensional subspace of the energy hypersurface.

[9]This time fraction may then also be regarded as the *probability* of finding the system in the macrostate corresponding to A: $P(A) \propto \mu(A)$.

apart from multiplicative and additive constants, we may define $S(t) = \log \mu[A(t)]$.[10]

Statistical mechanics now introduces an ensemble of identical systems on the same energy hypersurface and asks for the behavior of the function $S(t)$, averaged over the ensemble. According to Boltzmann's H-theorem, the ensemble-averaged value of $S(t)$ will necessarily increase with time until it has reached its maximum value, at which point we say the system has arrived at its equilibrium state. For individual systems, this means that the entropy is extremely likely to increase with time until the system has reached equilibrium. That's the thermodynamic arrow of time, applicable to every sufficiently large,[11] isolated physical system.

There were two famous kinds of objections to the H-theorem. The first is the kind we have already discussed earlier, based on the fact that the underlying dynamics of the particles making up the system is invariant under time reversal. Take a system that has developed, with growing entropy, to a state A near equilibrium; consider the same system in a state A' that differs from A only by having all the momenta of its constituents reversed, and let it go. Surely it will now retrace its trajectory in phase space back to where it started and its entropy will decrease, apparently contradicting the second law of thermodynamics.

The second famous objection was based on Poincaré's recurrence theorem, which states that every spatially confined mechanical system will eventually return arbitrarily closely to its initial condition. On the face of it, it seems impossible to reconcile this theorem with the second law.

Before answering these two objections, let us clearly understand what, exactly, is happening to the entropy of an isolated system with many degrees of freedom. Take two adjacent rooms, one much warmer than the other, and at the time $t = 0$ open a door

---

[10] As we saw on the last page, the volume ratio between the smallest and the largest star of an N-particle system is $N!$. By Stirling's formula, this means that for a liter of a gas at room temperature and atmospheric pressure, the corresponding entropy difference is of the order of 10^{24} times Boltzmann's constant (which was not included in the simple definition of entropy given).

[11] This means, with many degrees of freedom.

between them. What accounts for the fact that after a while, both rooms will have the same temperature? At the initial time, the point describing the system in its Γ-space is located in a star that is relatively small—its entropy low—because, when the air molecules located in the hotter room have a higher average kinetic energy than those located in the cooler room, there are not nearly as many microstates available for exchange of molecules in the same star as there are in the star in which the average kinetic energy is the same in both rooms. (The principal reason for this discrepancy is that the star of the first situation contains many more parts in which several molecules find themselves in identical coarse grains of the particle phase space, and their exchange does not add to the volume of the star. Every time a molecule moves from a grain in the particle phase space occupied by n others to an unoccupied grain, the volume of the star grows by a factor of 2^n.) In the course of time, the system point will traverse many stars, and the entropy will correspondingly fluctuate wildly. On the average, however, it will spend more time in the larger stars, and when it arrives in one of the largest, corresponding to "all the molecules more or less equally distributed and with the same average kinetic energy in every macroscopic spatial region" it will spend a very long time there; it has reached an equilibrium state. True, according to Poincaré's recurrence theorem it will eventually have to return to a situation in which the two rooms have very different temperatures, but this "eventually" is enormously long compared to the relaxation time of the system—longer, in fact, than the history of the universe—and so it plays no practical role.

The thermodynamic arrow of time unambiguously points from the instant of opening the door between the two rooms to the time at which both rooms are essentially at the same temperature. The important thing to remember is that the system started in a relatively small star and therefore with low entropy. Since the entropy is lowest when the system passes through the smallest stars, which it rarely visits, the farther a fluctuation deviates from the maximal entropy, the more rarely it occurs. Eventually, the system will reach one of the very large stars, not only to linger there for a long time—according to the ergodic theorem—but most likely to depart, when it does, for another large star that is almost indistinguishable from the first. Fluctuations in its entropy will still occur, though large ones will happen very rarely, with long intervening intervals.

But, notice that if we plot the entropy as a function of time, assuming that the initial location of the system in a small star (and hence with low entropy) occurred simply as a natural fluctuation, and we look at the graph for $t < 0$, everything I described above for positive times will hold equally well for negative times! In other words, in all its essential features, though not necessarily in detail, the entropy plot is *symmetric* about $t = 0$, and the system approaches equilibrium—the entropy approaches its maximal value—not only in the future but also in the past. The arrow of time has disappeared. The first "objection" to the second law, stated earlier, based on a reversal of the motion of all constituent molecules, is thus quite correct. For the reason elaborated upon below, however, it carries little force.

What should be clear from the discussion is that the origin of the thermodynamic arrow lies in the fact that certain system configurations, while extremely rarely occurring on their own as natural fluctuations, are easy to set up with extraneous assistance, and are the very configurations we usually postulate as initial conditions. It is not hard to arrange for two adjacent rooms to be at different temperatures and to open a door between them. If we were to wait for this situation to occur on its own, with both rooms initially at the same temperature, as a natural fluctuation, we would have to wait a very, very long time.[12]

However, statistical mechanics still has not provided us with an arrow of time, since, for a system left to its own devices, the entropy, if initially low, is seen to increase in both directions, toward the future as well as toward the past. It is the choice expressed in the way our questions are *always* posed—what happens *after*

---

[12]Similarly with the piece of sculpture and the new car: we can produce both by extraneous intervention in a relatively short time, but we would have to wait an eon for the junk to turn itself into a new vehicle or for the marble torso to become Venus on its own.

I open the door between the two rooms, quite analogous to the use of *initial* conditions for differential equations—which dictates a specific direction of time. Again, the reason the question is invariably asked in that form is our causal sense: we are looking for the *effect* of our act of opening the door, and this effect is expected later than our action. Opening the closed door is under our control; to ask for the development that might have naturally led to the low-entropy situation without the door being closed cannot be answered by causal means.

The essential symmetry of the entropy graph of an isolated system about any point of low entropy implies that *no matter which way the causal-cognitive arrow points, the direction of the thermodynamic arrow, pointing toward increasing entropy, will always be toward the future.*[13] As in the case of Boltzmann's H-theorem, one might argue that the way the problem is framed begs the entropy question, but the fair reply is simply that the thermodynamic and the causal arrow are not independent.

Thus we have come to two conclusions: (1) the three time arrows—the causal, the cognitive, and the thermodynamic arrow—are all connected in the sense of necessarily pointing in the same direction; and, as a consequence (2), the thermodynamic arrows of all isolated systems—the universe as a whole being one of them—agree with one another.

THE RADIATION ARROW

Let us now look briefly at what might be called arrow of time 3b: the direction of time in which radiation is found always flowing *away* from accelerated charges.

The Maxwell equations are symmetric under time reversal in the sense that if we change t into $-t$ while also reversing all cur-

---

[13] It is interesting to note the parallel between this time arrow, claimed to arise from thermodynamics, and the prohibition of superluminal signals, which is often said to arise from the theory of relativity. Both the thermodynamic arrow and the prohibition of signals faster than light actually rest, in addition, in an essential way on causality.

rents and the sign of the magnetic field, they remain unchanged. Nevertheless, in agreement with experimental results, radiation is predicted to be *emitted* by accelerated charges, flowing away rather than toward them, thus defining a *time arrow of radiation*.

This electromagnetic sense of time, however, is directly related to the thermodynamic arrow, which is why I called it arrow 3b. The detailed conditions of the field (including phases) may be considered analogous to the microstates of a mechanical system with (infinitely) many degrees of freedom, and the measurable field strengths on a surface analogous to a macrostate or coarse grain. The fields determined by initial conditions appropriate to "emission" by a point charge will later, at far distance, appear in the coarse grains of large spherical surfaces surrounding the charge; but, just as in the corresponding case of particle mechanics, they will fill these stars quite sparsely. In order to reverse the flow of radiation and send it back converging to a point, detailed phase conditions would have to be met—microstates specified—which would be practically impossible to fulfill: hence the irreversibility of the radiation flow. Still, this is not yet an arrow of time.

Recall how the arrow of *emission* of radiation comes about. The Maxwell equations with given charge and current distributions have infinitely many solutions, and we usually pick out one of these, by a specific choice of Green's function, as the only solution of physical interest: we use the *retarded* Green's function to define the *retarded* solution. This choice is dictated by the requirement that before the extraneous introduction of a point source in empty space, the electromagnetic field should vanish everywhere; only *after* the appearance of the point source should there be a field. Clearly, at the bottom of the choice of the retarded solution of the Maxwell equations as the only physically acceptable one lies the notion that it is the electric charge that *causes* the field; therefore, the field at a distant point depends on the behavior of the charge at an earlier time. From a purely mathematical point of view, the advanced solution would serve equally well, and so would the solution generated by a time-symmetric Green's function (or any one of infinitely many others).[14] Once this choice has been made, the radiation arrow of time has been fixed in the theory, accelerated charges are predicted to radiate away energy, and the time-reversal invariance of the underlying equations has been broken by the solutions we choose to accept as physically relevant. In other words, the direction of the radiation arrow of time, like its closely related thermodynamic counterpart, is ultimately determined by the causal arrow.[15]

THE COSMOLOGICAL ARROW

The fourth of the distinctly noticeable arrows of time in nature is determined by the observed expansion of the universe, pointing from a time when the universe was smaller to a time when it is larger. In other words, it is an arrow based on a specific observed fact, not on a theoretical structure or law. Again, the underlying differential equations of the general theory of relativity are invariant under time reversal, but the solution that describes the actually observed mass distribution—so far as we know—of the cosmos, and the course of its development, singles out expansion over contraction in the present epoch. The universe, we believe, started out with a big bang at a point, and it has been expanding ever since. (Whether this expansion will last forever, will stop, or will reverse itself is a separate question that has no direct bearing on the cosmological arrow of time as it exists now.) I don't believe there is anything puzzling about this, and the existence of the cosmological arrow, by itself, requires no further explanation.

[14] The absorber theory of Feynman and Wheeler was an attempt to do just that. For good or ill, it never caught on simply because it violated physicists' causal sense.

[15] Two papers, P. C. Aichelburg and R. Beig, "Radiation damping as an initial value problem, *Annals of Physics* **98** (1976), pp. 264–283, and J. L. Anderson, "Why we use retarded potential," *American Journal of Physics* **60** (1992), pp. 465–476, argue that it is the determination of the field from initial values (together with a condition of finite field energy), rather than causality, that leads to the choice of retarded solutions in radiation theory. As I have pointed out, the choice of *initial* values for solutions of field equations is also determined by causality.

161

However, there are physicists, notably Thomas Gold, who have argued that there is an intrinsic connection between the cosmological arrow of time and the thermodynamic arrow. Gold's reasoning is based, in the first instance, on noting that a system of many degrees of freedom, left completely isolated for a very long time, will tend to be in equilibrium and its entropy close to its maximal value; it will then have no thermodynamic arrow. External interference from a larger system, far from equilibrium, is required to produce a condition that introduces an arrow of time: "We have to go to a larger scale to understand how it contrived to know the arrow of time."[16] By this interference, Gold contends, the system *inherits* the direction of its thermodynamic arrow of time from the larger one.

The largest system from which to inherit the arrow, of course, is finally the universe. So Gold imagines a star inside an insulating box. After a very long time, the system inside the box will have reached equilibrium and thus time's arrow will be missing. Suppose now we open, for a moment, a small hole in the side of the box: the only effect will be that some radiation escapes and a very much smaller amount will enter. The reason for this asymmetry is that the universe outside the box has an almost unlimited capacity for absorbing radiation rather than reflecting it back, primarily because it is expanding. The sky is black owing to the fact that "in most directions the material on a line of sight is receding very fast, and its radiation, therefore, shifted very far to the red"[17] (the resolution of Olber's paradox).[18] In this manner, Gold sees the expansion of the universe as responsible for the radiation arrow, and the radiation arrow as the mechanism by which the box with the star in it "contrives to know" its thermodynamic arrow from the cosmological arrow, keeping it until it again reaches equilibrium.

It seems to me this argument is erroneous. First of all, if, for one reason or another, as Schulman points out,[19] more radiation en-

[16] T. Gold, "The arrow of time," *Americal Journal of Physics* **30** (1962), pp. 403–410.
[17] Ibid., p. 406
[18] At the time Gold wrote this, he believed in the steady-state universe, in which the darkness of the sky was entirely attributed to the expansion.
[19] Schulman, *Time's Arrow and Quantum Measurement*, pp. 126ff.

tered the box than escaped, not the other way around as Gold assumed, the equilibrium inside would still be disturbed and the entropy momentarily reduced. Therefore, after the opening is closed, the entropy would rise again and the thermodynamic arrow would point in the same direction as it did in the situation envisaged by Gold. While it is true that an outside influence usually sets the starting point of the course of events during which the entropy of an otherwise isolated system increases—alternatively, a system like the universe may simply start with an initial condition of low entropy—there is no reason to suppose the system in question *inherits* the direction of its time arrow from the larger outside system. The disturbance merely serves to produce a state far from equilibrium. No matter in which direction the causal—and hence cognitive—arrow then points, the system's entropy will increase in that direction. The expansion of the universe has the effect of causing or contributing to its own disequilibrium and low entropy;[20] its thermodynamic arrow then points in the direction determined by the expansion, true enough. But we cannot hold this responsible for the entropy increase of isolated subsystems of the universe.

The contention that the cosmological time arrow determines the direction of the thermodynamic one is therefore unfounded; the arrows are, so far as I can see, independent of each other. If it should turn out that eventually the universe begins to contract rather than expand forever, the world will experience in all respects the same time direction as now. On the other hand, the three arrows discussed earlier—the causal, the cognitive, and the thermodynamic—necessarily point in the same direction, toward a future determined ultimately by the fact that effects *follow after* their causes. If and when there is a future phase during which the universe grows smaller, the change will, in fact, be experienced as a contraction.[21]

At the center of the mystery of the time arrows, then, lies the simple, naked, observed fact of the universality of the temporal

[20] Provided the expansion is faster than the relaxation rate; see the discussion in ibid., pp. 132ff.

[21] If Gold were right, of course, a contraction of the universe could never be experienced.

163

order of cause and effect. It rests, let's remember, on a definition of the distinction between cause and effect based on *control* over the former. This control is exerted, in principle, by means of random devices, that is, "uncaused" events, and experimental physics is founded on the premise that such control is possible. But why does nature choose to arrange all effects to occur in the same time-order with respect to their causes?

To ask this question, now that we know that at bottom nature is ruled by quantum physics, may seem perverse. It should be remembered, though, that, while quantum mechanics is acausal when it comes to individual events, its strict probabilistic predictions allow us to distinguish statistically produced effects from their causes no less clearly than classical physics does invariable cause-effect connections. Quantum or no quantum, after sufficiently long experimentation, no physicist has ever been confused about whether A causes B or *vice versa*. And here, too, effects have never been seen to precede their causes.[22]

The reason nature arranges no causes to occur after their effects has to be sought in the existence of order. If some causes happened before their effects and some after, causal cycles of the kind mentioned in the last chapter would be possible: a causal chain of events could be set up that would prevent the occurrence of the very cause that was assumed to start it. The existence of such self-contradictory causal cycles would inevitably lead to chaos (in a much more serious sense than envisioned in the currently fashionable studies of "deterministic chaos"). In other words, nature without a universal causal arrow of time could not be orderly. We have, at this point, no other explanation.

---

[22] As mentioned in the previous chapter, there was a short period of confusion some thirty years ago when some physicists thought that quantum mechanics might provide a way out of the causality-based prohibition of signals faster than light and suggested that there might be *tachyons*, particles that travel faster than light. The idea was ill-founded and, though experimentally diligently sought, as I noted, tachyons were never found.

Quantum Mechanics and Reality

IN SEVERAL PREVIOUS chapters, we have already discussed various aspects of quantum mechanics, including some that appear to us strange and unintuitive: the difference between the definitions of the state of a physical system in classical and quantum physics, the probabilistic rather than causal nature of the latter—despite its deterministic character—and the consequences flowing from this. As I pointed out, many, but not all, of the counterintuitive characteristics of quantum mechanics follow simply from the fact that it is a probabilistic theory. In this chapter we are going to delve into those features which, for most people, cause the greatest unease and which incline even some of the greatest experts in the field, including Feynman, to feel that, at bottom, they "don't understand it."

SCHRÖDINGER'S CAT

The infamous case of Schrödinger's cat,[1] the *Gedanken* experiment Schrödinger constructed to show what he found impossible to accept about quantum mechanics, is a good place to start. He envisioned a cat imprisoned in a box, together with a closed glass vial filled with a lethal gas. A hammer, poised to shatter the vial, would be triggered by a mechanism activated by the decay of a radioactive atom with a half-life of an hour. According to quantum mechanics, he argued, the state of the cat after one hour had to be in a superposition of half dead and half alive. Only after the box was opened for inspection could the poor cat be either definitely dead or happily alive, a description of the world he regarded as

---

[1] E. Schrödinger, *Naturwissenschaften* **23** (1935) pp. 807–812, 823–828, 844–849; translation in J. A. Wheeler and W. H. Zurek, *Quantum Theory and Measurement* (Princeton: Princeton University Press, 1983).

intolerable; his *Gedanken* experiment has been echoed as an attack on quantum physics by many people ever since.

The effectiveness of Schrödinger's example, of course, rests first and foremost on his use of a living creature, which surely could not be in a state of suspended animation. Our initial reply to him therefore takes issue with the very premise of the argument, that a physical system as complicated as a large living organism, in constant interaction with its environment via its breathing, metabolism, etc., together with a radioactive atom, could possibly be in a quantum-mechanical pure state, described by a wave function. This aspect of the nicely invented *Gedanken* experiment is, of course, totally unrealistic, but we might try to reduce it to a physical situation unencumbered by emotional prejudices concerning superpositions of states of living organisms, and in such a form the conundrum still exists. It can then, however, be subjected to experimental tests.

Such an experiment was performed in 1996 at the National Institute of Standards and Technology.[2] By means of newly perfected techniques of laser cooling, Monroe and his coworkers were able to trap a single beryllium ion in two different positions, many atomic diameters apart, depending on whether its spin was up or down. The state of the ion they were able to produce was a *pure state*, made up of a superposition of *"position x_1 with spin up"* and *"position x_2 with spin down."* Only in response to a specific "inspection" would the beryllium ion "make up its mind" to be at position x_1 or at position x_2.

While an ion does not have quite the emotional impact of a cat, the principle here is the same; it does strike us as strange that we cannot assert definitely that an atom is *here* rather than *there*. The confusion, however, comes directly from our intuitive attachment to the classical notion of *state* as contrasted to the quantum-mechanical one, as discussed in chapter 2. Note that the essence of Schrödinger's objection does not really depend on the assumption that the cat exists in a *superposition* of two states, dead and alive; it would be applicable as well to any probabilistic theory.

---

[2]C. Monroe et al., "A 'Schrödinger cat' superposition state of an atom," *Science* **272** (May 24, 1996), pp. 1131–1136.

The only reason for choosing a quantum mechanical *pure* state was that such a state, he assumed, described an individual cat rather than an ensemble of cats. The proper resolution of the cat "paradox" is therefore that the meaning of "the state of a system" in quantum mechanics is not a direct description of reality.

THE EPR EXPERIMENT

This brings us straight to the famous debate we already briefly discussed in chapter 2 between Einstein, together with Podolsky and Rosen, on one side and Bohr on the other. Remember the more transparent form, given by David Bohm: a molecule of zero spin, consisting of two atoms, each of spin 1/2, in a singlet state, decays, with its constituents flying off in opposite directions. Since the total angular momentum of the two atoms is zero, their individual spin projections, no matter with respect to what direction, must be equal and opposite. This means that if we measure the x-projection of the spin of atom A and find the result positive, then we know, without ever having to approach and disturb atom B, that the x-projection of *its* spin must be negative. From this, EPR conclude that the x-projection of the spin of atom B corresponds to "an element of physical reality," which they define as follows: "If, without in any way disturbing a system, we can predict with certainty...the value of a physical quantity, then there exists an element of physical reality corresponding to this quantity."[3]

But we could just as well have measured the y-projection of the spin of atom A and thereby inferred the y-projection of the spin of atom B without disturbing it; thus the y-projection, too, corresponds to an element of physical reality. But, because the x and the y-projections do not commute, quantum mechanics does not allow us to determine both simultaneously; therefore, EPR conclude, "quantum mechanics has to be regarded as an incomplete description of reality."

[3] A. Einstein, P. Podolsky, and N. Rosen, "Can quantum mechanical description of physical reality be consider complete?" *Physical Review* **47** (1935), p. 777.

The reply of Bohr, who hated to disagree with Einstein, was, in its gist, as follows:

> The extent to which an unambiguous meaning can be attributed to such an expression as "physical reality" cannot of course be deduced from *a priori* philosophical considerations, but...must be founded on a direct appeal to experiments and measurements.... In fact, this new feature of natural philosophy means a radical revision of our attitude as regards physical reality.[4]

Whether we agree with Einstein or with Bohr on the question of reality, which we may leave in abeyance for now, there are some conclusions that may safely be drawn from EPR's argument. The first is that to ascribe the origin of Heisenberg's indeterminacy principle, which prevents us from measuring simultaneously the x and y-projections of the spin of a particle because the two don't commute,[5] to the disturbance of one by the measurement of the other, is untenable unless such a disturbance is transmitted by "spooky action at a distance," in Einstein's derisive phrase. The same "spooky action at a distance" may also by invoked to explain the strange phenomenon that if the x-projection of the spin of atom A is measured and found positive, a measurement of the x-projection of the spin of atom B, far away, must yield a negative result with certainty, but if the y-projection of the spin of A had been measured, a measurement of the x-projection of the spin of B would yield a negative result only with probability $1/2$—"How does it know?" As I pointed out by a simple example in chapter 6, however, this kind of strangeness adheres to any probabilistic theory and not just to quantum mechanics. Whenever the probabilities of two distant events are correlated, the outcome of a test on one must influence the probability of the other, and one may always be tempted to ask, *How does it know?*. This, however, does not exhaust the strangeness of the EPR-Bohm *Gedanken* experiment.

---

[4] N. Bohr, "Can quantum mechanical description of physical reality be considered complete?" *Physical Review* **48** (1935), p. 696.

[5] Actually, the original EPR paper uses positions and momenta rather than spin projections, so that the connection to the Heisenberg principle is more direct.

What is particularly quantum mechanical about the two parti
cles in the EPR case is that they are *entangled* in a way withou
analogue in other probabilistic particle theories. In fact, John Be]
proved there can be no local probabilistic theory that would al
ways give the same result as quantum mechanics, or even a result that approximated it. For the case discussed above, as well as
for all other correlations of particle variables measured at a non-
zero distance from one another, the difference between the joint
probabilities of the measured values of two variables (such as the
spin projections of the two particles in the EPR-Bohm *Gedanken* experiment) predicted by quantum mechanics on one hand and by
any classical probabilistic theory in which the two spatially separated measurements are independent of one another—the assumption that the theory is *local*—on the other hand, is always greater
than some positive number (which in the case discussed above
is $[\sqrt{2} - 1]/4$).[6] This result is known as *Bell's inequality*, and the
consequence, that no local "hidden variable theory" could duplicate all the probabilistic predictions of quantum mechanics, is *Bell's
theorem.*

Going beyond ordinary probability correlations, quantum entanglement is a matter of *phases* of wave functions and hence ultimately rests on the wave-particle duality. It leads to mutual depen-
dencies of particles at large distances from one another for which
we have no intuitive understanding and which are sometimes interpreted as giving quantum mechanics a *nonlocal* character. However, since quantum mechanics attaches a wave aspect to all entities called particles, the term "nonlocal," which for particles has
derogatory connotations—such as *spooky*—is not quite apposite.
Waves are by their very nature nonlocal, and we have no trouble
understanding long-range phase correlations among them. (Think,
for example, of the two-slit Young experiment or of large-scale
interferrometry.) It is only in the particle language that our intuition balks. Whatever nonlocality is intrinsic to quantum mechan-

[6]J. S. Bell, "On the Einstein Podolsky Rosen Paradox," *Physics (N.Y.)* **1** (1964),
pp. 195–200.

CHAPTER 8

ics originates in the well-established fact that all particles have a wave nature as well.

By contrast, Bell's theorem implies that any hidden-variable theory, such as the one proposed by Bohm, which would circumvent some of the "weird" aspects of quantum mechanics—without disagreeing with any of its experimental predictions—by means of an underlying causal substratum consisting of unobservable particles, would necessarily have to be nonlocal in a fundamental way. Since such theories contain no waves as essential ingredients, their nonlocality is a much more serious introduction of spooky action at a distance than any quantum mechanical nonlocality originating from its waves. (Faraday, remember, introduced fields for the principal purpose of avoiding Newton's action at a distance!) If the historical development had been such that a hidden-variable theory was first to be accepted to account for all the newly observed phenomena that cried out for explanations in the 1920s, the later introduction of quantum-mechanics might well have been regarded as a welcome novel theory that avoided both the introduction of physically unobservable particles and the obnoxious nonlocal action at a distance to which they were subject.

BELL EXPERIMENTS

The Bell inequality brought the discussions centered around EPR down to earth. For years the debate had remained at a more or less philosophical level, most physicists siding with Bohr (not that they really bothered to try to understand him—Bohr often tended to sound like an oracle). It was John Bell's inequality that focused the problem to a point where it could be subjected to a clear-cut experimental test. We'll use another *Gedanken* experiment to clarify Bell's idea.

The setup consists of a transmitter T that sends two correlated messages—by means of particles with appropriate properties or by any other means—to two distant receivers, A and B, in opposite

Receiver A

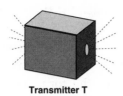

Transmitter T

Receiver B

Figure 8.1. Signals are sent from the transmitter T to the two receivers A and B, each of which have a red and a green light and can be set at either 1 or 2.

directions from T.[7] Each receiver has two settings, 1 and 2, and two colored lights, one red, one green (see fig. 8.1). In each run of the experiment a pair of messages is sent to the two receivers, which, independently and randomly, have been set at either 1 or 2, and upon receipt of their messages, flash either their green or their red lights. Great care has been taken to assure that the two receivers are not in communication with each other, that neither is aware of the message sent to the other, and that the messages sent are not influenced by the settings of the receivers. The individual results are recorded in the form, say, 2G1R, meaning "receiver A, set at 2, flashes green, and receiver B, set at 1, flashes red." When the results of a long sequence of runs are examined, the data are found to have the following characteristics:

a. In no case in which the two settings of A and B differed did both flash green; in other words, neither 1G2G nor 2G1G ever occurred.

b. When both A and B were set at 1, at least one of them flashed green; i.e., 1R1R never happened.

c. When both A and B were set at 2, one-third of the time both flashed green, i.e., 2G2G happened. (That 2G2G happened exactly one-third of the time is not really important; what matters is that it occurred at all.)

---

[7] My description is based on N. David Mermin, "Quantum mysteries refined," which, in turn, is based on Lucien Hardy, "Non-locality for two particles without inequalities for almost all entangled states," *Physical Review Letters* **71** (1993), pp. 731–734. See also Mermin's earlier papers in the *American Journal of Physics* **49** (1981), pp. 940–943, and **58** (1990), pp. 731–734.

The question is how to account for these features by imagining what, in terms of "real" signals, the dual messages could possibly have been. Let us denote a pair of messages by, say, $[GR, RR]$, meaning "signal to A: if set at 1, flash green, if set at 2, flash red; signal to B: if set at 1, flash red, if set at 2, flash red." Now, (a) implies that no messages of the form $[G \cdot, \cdot G]$ or of the form $[\cdot G, G \cdot]$ were sent, and (b) implies that no messages of the form $[R \cdot, R \cdot]$ were sent, either. But 2G2G, which according to (c) did happen, can be brought about only by one of the four messages of the form $[\cdot G, \cdot G]$, each of which would lead to a contradiction with (a) or (b): the four are $[RG, RG]$, $[GG, RG]$, $[RG, GG]$, and $[GG, GG]$; the first would violate (b), and the other three would violate (a). We conclude that the data, when interpreted in terms of ordinary messages and without "spooky action-at-a-distance" communication between the receivers, are logically contradictory. Specifically, the restrictions on correlations implied by (a) and (b) are inconsistent with (c). They can, nevertheless, be produced by means of signals utilizing quantum particles.

Here is the way it can be done. Assume that the transmitter contains unstable molecules of spin 1 in the following superposition of states of spin projections 0 and -1 with respect to the z-axis:

$$\Psi = \frac{1}{\sqrt{3}}[\sqrt{2}\Psi_0 + i\Psi_{-1}],$$

and each of the molecules is made up of two atoms of spin $1/2$, so that its spin state is given by

$$\Psi = \frac{1}{\sqrt{3}}[\psi_+ \otimes \psi_- + \psi_- \otimes \psi_+ + i\psi_- \otimes \psi_-]$$

in terms of the spin *up* and *down* states of its constituent atoms. The two receivers A and B have detectors for the decay products with Stern-Gerlach devices oriented vertically if set at 1 and horizontally if set at 2; if spin *up* or spin *right* is found, the red light flashes, and if spin *down* or spin *left* is found, the green light blinks.

Feature (b) is now immediately obvious from the fact that Ψ is orthogonal to $\psi_+ \otimes \psi_+$. Using[8] $(\psi_r, \psi_+) = (\psi_l, \psi_-) = \frac{1+i}{\sqrt{2}}$ and

---

[8] We denote the two possible values of the horizontal spin projection by l and r.

$(\psi_l, \psi_+) = (\psi_r, \psi_-) = \frac{1-i}{\sqrt{2}}$, we find that

$$(\psi_l \otimes \psi_-, \Psi) = \frac{1}{\sqrt{3}}[(\psi_l, \psi_+) + i(\psi_l, \psi_-)] = 0,$$

and similarly, $(\psi_- \otimes \psi_l, \Psi) = 0$, which means that feature (a) holds. It is finally a matter of a simple calculation to ascertain that $(\psi_l \otimes \psi_l, \Psi) = \frac{1}{\sqrt{3}}$, which implies (c).[9]

In principle, one could imagine carrying out this experiment and checking whether the result really agrees with quantum mechanics. This turned out to be too difficult. There are, however, other experiments along these lines that are feasible and have in fact been carried out.[10] Three of the type suggested by the EPR-Bohm argument but with linear polarization correlations of pairs of photons emitted in a radiative cascade of calcium atoms instead of spin correlations of particles of spin 1/2 were performed by Alain Aspect and various collaborators.[11] All of these tests confirmed the predictions of quantum mechanics and, via Bell's inequalities, ruled out the possibility of accounting for the results by means of any kind of local "realistic" theory.[12]

WHAT SETS QUANTUM MECHANICS APART?

So what are the special characteristics of quantum mechanics that set it apart from "classical physics"? First and foremost, we have to

---

[9] The Mermin paper is more general than the specific case given here and contains further discussion.

[10] For a general discussion of such tests and their implications, see J. F. Clauser and A. Shimony, "Bell's theorem: Experimental tests and implications," *Rep. Progr. Phys.* **41** (1978), p. 1881.

[11] A. Aspect, P. Grangier, and R. Roger, "Experimental tests of realistic local theories via Bell's theorem," *Physical Review Letters* **47** (1981), p. 460, and "Experimental realization of Einstein-Podolsky-Rosen-Bohm *Gedankenexperiment*: A new violation of Bell's inequalities," ibid. **49** (1982), p. 91; A. Aspect, J. Dalibard, and G. Roger, "Experimental tests of Bell's inequalities using time-varying analyzers," ibid. **49** (1982), p. 1804.

[12] They have, however, not yet convinced everybody. Occasional arguments about the reliability of the experiments still break out.

mention that it accounts for the existence of identical, stable "particle" states such as nuclei, atoms, and molecules, which are totally foreign to classical physics. Bohr's contribution of intuitively introducing these states was seminal, sowing the seeds for the ideas of Heisenberg, Schrödinger, and Dirac. This remarkable contribution, however, came at the price of introducing not only an acausal theory, but one that contains elements even more counterintuitive than probabilistic correlations. In chapter 6 we discussed the "collapse of the wave function" and other "paradoxes" associated with the quantum theory of measurement, and learned that any probabilistic theory would contain analogous effects. Though these are often cited as the quantum world's most peculiar properties, they are strange only because our intuition is slow to adapt itself to probabilistic theories. The wave function collapse becomes truly *spooky* only if we make the mistake of thinking of that function as a real condition of physical space, so that the collapse is an observable, instantaneous effect over long distances. The state vector, remember, is the abstract quantum-mechanical representation of the state of a system—it is not a direct reflection of reality— and its *representor*, the wave function, lives not in physical space but in configuration space, which, for an N-particle system, is $3N$-dimensional. Even for a single particle, for which the configuration space has the same dimensionality as physical space, these spaces are not identical and must not be confused.

The legitimately counterintuitive aspect of quantum mechanics —there are many that seem undeniably strange—all have to do with the superposition principle and the concomitant entanglements of particles; these are phase correlations ultimately based on the fact that the wave-particle duality prevents quantum-mechanical "particles" from being real particles in the sense of the word as it is used in the macroworld.

In most instances, what high-energy physicists have in mind when they announce the discovery of a new "particle" in a big scattering experiment, is that they have found a "resonance," a noticeable sharp bump in the plot of a reaction or collision cross section as a function of the energy of the primaries. Such a sharp local increase of the cross section indicates that the colliding particles were able to form, temporarily, a more-or-less localized state—the new, unstable particle—which subsequently decays. The mass of the new particle is related, by the relativistic energy-momentum relation, to the energy at which the peak is centered, and its lifetime τ is obtained from the width Γ of that peak as $\tau = \hbar/\Gamma$. Thus, if the bump is very sharp relative to the structure of the background, the evidence for a particle is clear and convincing; if it is broad, on the other hand, it is intrinsically ambiguous, especially if you remember that all experimental data have errors attached to them. Therefore, the dividing line between the existence or nonexistence of a particle is quite fuzzy.

Just how ambiguous the evidence for the existence of particles can be was demonstrated by the case of the purported discovery in 1967 that a "particle" called the A_2 might actually be two particles. A group of experimenters at CERN had found that their data on reactions of the type $\pi^- p \to p X^-$ seemed to indicate that a resonance called the "A_2 meson" had an unusual shape: the resonance peak appeared to be split in two, with a deep dip in the middle, indicating the existence of *two* particles of almost the same mass.[13] Another group, from Northeastern University, performing a similar experiment, failed to see the dip at the top of the peak, even though they tried their best to find it in their data. The controversy was publicly fought out in 1971 at a meeting of the American Physical Society, with the CERN group arguing that the Northeastern group did not find the split because their instruments were simply not as sensitive as those at CERN. However, as more data accumulated—and, in addition, the CERN group realized that they had made a basic error in the handling of their data[14]—the dip finally disappeared and the "A_2 meson" remained a single, undivided "particle." Even without the data-handling mistake in this instance, it is clear that in the course of accumulating statistical

---

[13] G. Chicovani et al., "Evidence for a two-peak structure in the A_2 meson," *Physics Letters* **25B** (1967), p. 44.

[14] They had treated batches of data that showed the dip differently from batches that did not, always finding something wrong among the latter when carefully scrutinized; see A. Cromer, *Uncommon Sense* (Oxford: Oxford University Press, 1993), pp. 169–170.

evidence for the existence of a particle, based on scattering data indicating an apparent resonance, which is all we ever have for many of them, conclusions that are spurious and evanescent are bound to be drawn from time to time. What is more, when the bump visible in the data is relatively broad, the evidence for the existence of the alleged "particle" may never be more than dim.

For these reasons, it seems to me that we should eliminate particles as the fundamental entities from which to form physical theories at the submicroscopic level and in terms of which to think of reality. This does not mean that, like Ernst Mach, I don't believe in atoms. At a certain level of description, molecules, protons, neutrons, electrons, and all the other building blocks of matter, surely exist; the overwhelming evidence for their reality cannot be denied. But whatever we mean by the emotionally charged term of *reality*, whose existence we, as physicists, cannot seriously doubt, has to be described on two scales—the large scale that includes everyday experience, ranging from the galactic down to the microscopic (in the literal sense of the word), and the submicro scale, far below the reach of our senses and any direct acquaintance. This is the scale of the quantum world, a world generally hidden from our immediate experience, though it sometimes "peeks through," as in superconductivity, and makes itself felt directly even on the macro scale.

The submicro reality should not be confused with what philosophers call *ultimate reality*; whatever that phrase signifies, it is not the subject of physics. But the difficulty we face as physicists investigating the submicro world is that we have to describe what we find, we have to communicate the explanations of our findings; private, secret science is an oxymoron. One language available to us for this purpose is that of mathematics, but this language is not accessible to the understanding of most nonscientists or natural to the intuitive sense of many physicists.

The only other language we have is based on concepts necessarily structured to accommodate our experiences at the larger, everyday scale, which is the reason why Heisenberg insisted that "the words of this language represent the concepts of classical physics.... Therefore, any statement about what has 'actually happened' is a statement in terms of classical concepts."[15] *This language, however, is inadequate.* Much of the perceived "weirdness" of the quantum world is the result of such an inadequacy; in other words, it resides in the linguistic order rather than in the reality of that world. When Bohr proclaimed, "There is no quantum world. There is only an abstract quantum mechanical description,"[16] this is, I believe, at least in part what he meant. But to blame our difficulties in understanding the quantum world on our "linguistic order" is not meant to downgrade them; after all, our language and concepts based on it are all we have at our disposal to construct an image of reality. We can imagine only what fits the screen of our monitor.

However, Bohr's philosophical position (idealism)—"There is no quantum world"—goes farther than is justified by what we know. An adequate view of the reality of the "quantum world," at the submicroscopic scale, is possible, but neither in terms of particles nor of waves; rather, it must be constituted in terms of quantum fields, "perhaps," in the words of Julian Schwinger,[17] "the deepest expression of what has been learned within the framework of microscopic phenomena." The field, in turn, generates "particles," as we have seen in chapter 4, and their states are described quantum mechanically by means of wave functions—or simply "waves"— the two concepts of particles and waves thus being inextricably intertwined by a duality that generates paradoxes whenever the former are mistakenly imagined as little billiard balls or the latter as analogues of classical electric fields. Thus, both particles

---

[15] Werner Heisenberg, *Physics and Philosophy* (New York: Harper, 1958), p. 144. This is also the reason why Heisenberg and Bohr emphasized that experimental results always had to be expressed in terms of classical physics. Some physicists find this "encapsulating" of quantum mechanics within classical physics very unnatural and stress that quantum mechanics should be able to stand on its own feet, with the classical results emerging from it in the appropriate limit (which is not always simply and straight-forwardly $\hbar \to 0$). Of course, it does and they do, but the linguistic difficulties in communication are there nonetheless.

[16] Quoted by Age Petersen, "The philosophy of Niels Bohr," *Bulletin of the Atomic Scientists* **19** (September 1963), p. 12.

[17] Quoted in S. S. Schweber, *QED and the Men Who Made It: Dyson, Feynman, Schwinger, and Tomonaga* (Princeton: Princeton University Press, 1994), p. 357, from p. 14 of an unpublished manuscript by Schwinger.

and waves are epiphenomena rather than fundamental entities of nature.

Particles are not creations of our minds, but they are the fundamental building blocks of matter, as the ancient Greek atomists took them to be, only when viewed at the upper scale of reality; at the submicro scale, they dissolve. First the molecules dissolve into atoms, then the atoms dissolve into electrons and a nucleus, then the nucleus into nucleons, followed by the dissolution of nucleons into quarks, and who knows what comes after that. At the submicro scale, all this graininess and corpuscularity becomes a surface phenomenon produced by the quantum field whence it originates. As I have tried to make clear, if we begin physics with what appears like classical particles and their degrees of freedom, quantum mechanics produces waves associated with them, waves that are needed for their representation; if we start with what seems classically like waves, either in the form of fields or via coherent vibrations of infinite lattices (as in solids), quantum mechanics generates particles with nonclassical properties. The closest we can come to submicroscopic reality is the quantum field; "particles" appear as its quantum manifestations, and "waves" are needed to describe the behavior of the particles. The origin of their apparently strange actions has to be sought in our attempts to describe these secondary phenomena and their behavior, using an inadequate language.

No doubt the most appropriate language by which to describe the quantum world is that of mathematics, which presents neither puzzles nor paradoxes, and with which experimental predictions can be made without ambiguities. But such an abstract method of communication is insufficiently concrete, especially for honing the intuition of experimenters and for conceptualizing the physical world. As physicists rather than mathematicians, whose fundamental purposes and aims, after all, differ from ours, we have to put up with the awkward consequences of using a language and conceptual machinery originating in the macroworld, and, alas, they are not up to the task.

Epilogue

WHEN WE LOOK over the whole of physics, that enormous domain of intellectual activity comprising a great diversity of creative approaches and styles as well as a wealth of facts, are we able to discern any salient characteristics of our view of nature at the beginning of the 21st century? With the due caution such a question deserves, I would identify three broad elements: at the most basic level, nature is best described in terms of the quantum field (which may eventually include a quantum geometry of space-time, but that remains to be seen); the explanatory machinery includes causality tempered by a pervasive appearance of probability; and the only language capable of describing nature unambiguously— the most efficient tool we have for thinking about it as well— is mathematics. While the latter two elements are of older vintage, the first is a new contribution of the 20th century. Although we may expect many new details to be added in the future, it seems unlikely that these features will undergo any fundamental changes, but then, our descendants could be in for a surprise as jolting as the quantum field would have been for Maxwell. Whatever awaits the next generations, we are not, as some speculate, approaching the end of physics!

Further Reading

Introduction and General

Barrow, John D., and Frank J. Tipler. *The Anthropic Cosmological Principle.* Oxford: Oxford University Press, 1986.

Dirac, P.A.M. "The cosmological constants." *Nature* **139** (20 February 1937), pp. 323–324.

Duhem, Pierre. *The Aim and Structure of Physical Theory.* Princeton: Princeton University Press, 1991.

Einstein, Albert. *Ideas and Opinions* [translation of *Mein Weltbild*]. New York: Crown Publishers, 1954.

Einstein, Albert. *Out of My Later Years.* New York: Philosophical Library, 1950.

Giere, Ronald. *Explaining Science.* Chicago: University of Chicago Press, 1988.

Heisenberg, Werner. *Physics and Philosophy: The Revolution in Modern Science.* New York: Harper, 1958.

Holton, Gerald. *Einstein, History, and Other Passions: The Rebellion against Science at the End of the Twentieth Century.* Reading, Mass.: Addison-Wesley, 1996.

Kuhn, Thomas S. *The Structure of Scientific Revolutions.* Chicago: University of Chicago Press, 1970.

Margenau, Henry. *The Nature of Physical Reality: A Philosophy of Modern Physics.* New York: McGraw-Hill, 1950.

Nagel, Ernest. *The Structure of Science: Problems in the Logic of Scientific Explanation.* Indianapolis: Hackett, 1979.

Newton, Roger G. *The Truth of Science: Physical Theories and Reality.* Cambridge, Mass.: Harvard University Press, 1997.

Poincaré, Henri. *The Foundations of Science (Science and Hypothesis, The Value of Science, Science and Method).* Lancaster, Pa.: Science Press, 1946 (originally published in 1913).

Popper, Karl R. *The Logic of Scientific Discovery.* New York: Basic Books, 1959.

Popper, Karl R. *Quantum Theory and the Schism in Physics.* Totowa, N.J.: Rowan and Littlefield, 1982.

Popper, Karl R. *Realism and the Aim of Science,* from the *Postscript to the Logic of Scientific Discovery.* W. W. Bartley, III, ed. Totowa, N.J.: Rowman and Littlefield, 1983.

Webb, J. K. et al. "Search for time variation of the fine structure constant." *Physical Review Letters* **82** (1999), pp. 884–887.

Weinberg, Steven. *Dreams of a Final Theory*. New York: Pantheon Books, 1992.

Chapter 1
Theories

Duhem, Pierre. *The Aim and Structure of Physical Theory*. Princeton: Princeton University Press, 1991.

Einstein, Albert. *Out of My Later Years*. New York: Philosophical Library, 1950.

Giere, Ronald N. *Explaining Science*. Chicago: University of Chicago Press, 1988.

Lakatos, Imre. "Falsification and the methodology of scientific research programmes." In *Criticism and the Growth of Knowledge*, Imre Lakatos and Alan Musgrave, eds., pp. 91–195. Cambridge, U.K.: Cambridge University Press, 1970.

Laudan, Larry. *Science and Values: The Aims of Science and Their Role in Scientific Debate*. Berkeley: University of California Press, 1984.

Laudan, Larry. *Beyond Positivism and Relativism: Theory, Method, and Evidence*. Boulder, Colo.: Westview Press, 1996.

Nagel, Ernest. *The Structure of Science: Problems in the Logic of Scientific Explanation*. Indianapolis: Hackett, 1979.

Popper, Karl R. *The Logic of Scientific Discovery*. New York: Basic Books, 1959.

Chapter 2
The State of a Physical System

Jauch, Joseph M. *Foundations of Quantum Mechanics*, chapter 6. Reading, Mass.: Addison-Wesley, 1968.

Chapter 3
The Power of Mathematics

Dyson, Freeman. "Mathematics in the physical sciences." In *The Mathematical Sciences*, COSRIM, eds., pp. 97–115. Cambridge, Mass.: MIT Press, 1969.

Einstein, Albert. *Ideas and Opinions* [translation of *Mein Weltbild*]. New York: Crown Publishers, 1954.

Flake, Gary William. *The Computational Beauty of Nature: Computer Explorations of Fractals, Chaos, Complex Systems, and Adaptation.* Cambridge, Mass.: MIT Press, 1998.

Hadamard, Jacques. *The Psychology of Invention in the Mathematical Field.* Princeton: Princeton University Press, 1945.

Hardy, G. H. *A Mathematician's Apology.* Cambridge, U.K.: Cambridge University Press, 1993.

Poincaré, Henri. *The Foundations of Science (Science and Hypothesis, The Value of Science, Science and Method).* Lancaster, Pa.: Science Press, 1946 (originally published in 1913).

Wigner, Eugene. "The unreasonable effectiveness of mathematics in the natural sciences." *Communications in Pure and Applied Mathematics* **13** no. 1 (1960). Reprinted in *Symmetries and Reflection,* pp. 222–237. Bloomington: Indiana University Press, 1967.

CHAPTER 4

FIELDS AND PARTICLES

Duck, Ian., and E.C.G. Sudarshan. "Toward an understanding of the spin-statistics theorem." *American Journal of Physics* **66** (1998), pp. 284–303.

Schweber, Sylvan S. *QED and the Men Who Made It: Dyson, Feynman, Schwinger, and Tomonaga.* Princeton: Princeton University Press, 1994.

Teller, Paul. *An Interpretive Introduction to Quantum Field Theory.* Princeton: Princeton University Press, 1995.

Weinberg, Steven. *The Quantum Theory of Fields.* Cambridge, U.K.: Cambridge University Press, 1995.

CHAPTER 5

SYMMETRY IN PHYSICS

Cornwell, J. F. *Group Theory in Physics.* London: Academic Press, 1984.

Lichtenberg, D. B. *Unitary Symmetry and Elementary Particles.* New York: Academic Press, 1978.

Mills, Robert. "Gauge fields." *American Journal of Physics* **57** (1989), p. 493.

Rosen, Joe. *A Symmetry Primer for Scientists.* New York: John Wiley & Sons, 1983.

Wigner, E. P. *Group Theory.* New York: Academic Press, 1959.

CHAPTER 6
CAUSALITY AND PROBABILITY

Ballentine, L. E. "The statistical interpretation of quantum mechanics," *Reviews of Modern Physics* **42** (1970), pp. 358–381.

Hacking, Ian. *The Emergence of Probability: A Philosophical Study of Early Ideas about Probability, Induction and Statistical Inference.* Cambridge, U.K.: Cambridge University Press, 1975.

Home, D., and Whitaker, M. A. B. "Ensemble interpretations of quantum mechanics. A modern perspective." *Physics Reports* **210** (1992), pp. 223–317.

Kolmogorov, A. N. "On the logical foundations of probability theory." In *Lecture Notes in Mathematics* **1021** (1983), pp. 1–5.

Mises, Richard von. *Probability, Statistics, and Truth.* New York: Macmillan, 1957.

Plato, Jan von. *Creating Modern Probability: Its Mathematics, Physics, and Philosophy in Historical Perspective.* Cambridge, U.K.: Cambridge University Press, 1994.

Salmon, Wesley C. *Causality and Explanation.* Oxford: Oxford University Press, 1998.

CHAPTER 7
ARROWS OF TIME

Bayer, Hans Christian von. *Maxwell's Demon: Why Warmth Disperses and Time Passes.* New York: Random House, 1998.

Gold, Thomas. "The arrow of time." *American Journal of Physics* **30** (1962), pp. 403–410.

Halliwell, J. J., J. Pérez-Mercader, and W. H. Zurek. *Physical Origins of Time Asymmetry.* Cambridge, U.K.: Cambridge University Press, 1994.

Schulman, L. S. *Time's Arrow and Quantum Measurement.* Cambridge, U.K.: Cambridge University Press, 1997.

CHAPTER 8
QUANTUM MECHANICS AND REALITY

Ballentine, L. E. "The statistical interpretation of quantum mechanics." *Reviews of Modern Physics* **42** (1970), pp. 358–381.

Bell, J. S. "On the Einstein Podolsky Rosen Paradox." *Physics (N.Y.)* **1** (1964), pp. 195–200.

Bohr, Niels. "Can quantum mechanical description of physical reality be considered complete?" *Physical Review* **48** (1935), p. 696.

Cushing, James T. *Quantum Mechanics: Historical Contingency and the Copenhagen Hegemony.* Chicago: University of Chicago Press, 1994.

Einstein, A., B. Podolsky, and N. Rosen. "Can quantum mechanical description of physical reality be considered complete?" *Physical Review* **47** (1935), p. 777.

d'Espagnat, Bernard. *Veiled Reality: An Analysis of Present-Day Quantum Mechanical Concepts.* Reading, Mass.: Addison-Wesley, 1995.

Gell-Mann, Murray, and James B. Hartle. "Quantum mechanics in the light of quantum cosmology." In *Complexity, Entropy, and the Physics of Information,* W. H. Zurek, ed., pp. 425-458. Reading, Mass.: Addison-Wesley, 1991.

Heisenberg, Werner. *Physics and Beyond: Encounters and Conversations.* New York: Harper and Row, 1971.

Home, D., and M. A. B. Whitaker. "Ensemble interpretations of quantum mechanics. A modern perspective." *Physics Reports* **210** (1992), pp. 223–317.

Jammer, Max. *The Philosophy of Quantum Mechanics—The Interpretations of Quantum Mechanics in Historical Perspective.* New York: John Wiley & Sons, 1974.

Margenau, Henry. *The Nature of Physical Reality: A Philosophy of Modern Physics.* New York: McGraw-Hill, 1950

Margurger, John H. III. "What is a photon?" *Physics Teacher* **34** (1996), pp. 482–486.

Mermin, N. David. "Is the moon there when nobody looks? Reality and the quantum theory." *Physics Today,* April 1985, pp. 38–47.

Mermin, N. David. "Quantum mysteries refined." *American Journal of Physics* **62** (1994), pp. 880–887.

Omnès, Roland. *The Interpretation of Quantum Mechanics.* Princeton: Princeton University Press, 1994.

Bibliography

Abraham, Ralph, and Jerrold E. Marsden. *Foundations of Mechanics*. Reading, Mass.: Benjamin/Cummings, 1978.

Arnold, V. I. *Mathematical Methods of Classical Mechanics*. New York: Springer-Verlag, 1978.

Ballentine, L. E. "The statistical interpretation of quantum mechanics." *Reviews of Modern Physics* **42** (1970), pp. 358–381.

Barrow, John D. *Pi in the Sky: Counting, Thinking, and Being*. Oxford: Clarendon Press, 1992.

Barrow, John D., and Frank J. Tipler. *The Anthropic Cosmological Principle*. Oxford: Oxford University Press, 1986.

Bayer, Hans Christian von. *Maxwell's Demon: Why Warmth Disperses and Time Passes*. New York: Random House, 1998.

Bell, J. S. "On the Einstein Podolsky Rosen Paradox." *Physics (N.Y.)* **1** (1964), pp. 195–200.

Bohr, Niels. "Can quantum mechanical description of physical reality be considered complete?" *Physical Review* **48** (1935), p. 696.

Clauser, J. F., and A. Shimony, "Bell's theorem: Experimental tests and implications." *Rep. Progr. Phys.* **41** (1978), p. 1881.

Cornwell, J. F. *Group Theory in Physics*. London: Academic Press, 1984.

Cromer, A. *Uncommon Sense: The Heretical Nature of Science*. Oxford: Oxford University Press, 1993.

Cushing, James T. *Quantum Mechanics: Historical Contingency and the Copenhagen Hegemony*. Chicago: University of Chicago Press, 1994.

Davies, P.C.W. *The Physics of Time Asymmetry*. Berkeley: University of California Press, 1974.

d'Espagnat, Bernard. *Veiled Reality: An Analysis of Present-Day Quantum Mechanical Concepts*. Reading, MA: Addison-Wesley, 1995.

Dirac, P.A.M. "The cosmological constants." *Nature* **139** (20 February 1937), pp. 323-324.

Duck, Ian, and E.C.G. Sudarshan. "Toward an understanding of the spin-statistics theorem." *American Journal of Physics* **66** (1998), pp. 284–303.

Duhem, Pierre. *The Aim and Structure of Physical Theory*. Princeton: Princeton University Press, 1991.

Dyson, Freeman. "Mathematics in the physical sciences." In *The Mathematical Sciences*. COSRIM, eds., pp. 97–115. Cambridge, Mass.: MIT Press, 1969.

Einstein, Albert. *Ideas and Opinions* [translation of *Mein Weltbild*]. New York: Crown Publishers, 1954.

Einstein, Albert. *Out of My Later Years.* New York: Philosophical Library, 1950.

Einstein, A., B. Podolsky, and N. Rosen. "Can quantum mechanical description of physical reality be considered complete?" *Physical Review* **47** (1935), p. 777.

Everett, H., III. " 'Relative state' formulation of quantum mechanics." *Reviews of Modern Physics* **29** (1957), pp. 454–462.

Flake, Gary William. *The Computational Beauty of Nature: Computer Explorations of Fractals, Chaos, Complex Systems, and Adaptation,* chapter 3. Cambridge, Mass.: MIT Press, 1998.

Gell-Mann, Murray, and James B. Hartle. "Quantum mechanics in the light of quantum cosmology." In *Complexity, Entropy, and the Physics of Information.* W. H. Zurek, ed., pp. 425-458. Reading, Mass.: Addison-Wesley, 1991.

Giere, Ronald. *Explaining Science.* Chicago: University of Chicago Press, 1988.

Gold, Thomas. "The arrow of time." *American Journal of Physics* **30** (1962), pp. 403–410.

Griffiths, Robert B. "A consistent history approach to the logic of quantum mechanics." In *Symposium on the Foundations of Modern Physics,* K. V. Laurikainen, C. Montonen, and K. Sunnarborg, eds., pp. 85–97. Gif-sur-Yvette, France: Editions Frontières, 1994.

Griffiths, Robert B. "Consistent histories and quantum reasoning." *Physical Review A* **54** (1996), pp. 2759–2774.

Hacking, Ian. *The Emergence of Probability: A Philosophical Study of Early Ideas about Probability, Induction and Statistical Inference.* Cambridge, U.K.: Cambridge University Press, 1975.

Hadamard, Jacques. *The Psychology of Invention in the Mathematical Field.* Princeton: Princeton University Press, 1945.

Halliwell, J. J., J. Pérez-Mercader, and W. H. Zurek. *Physical Origins of Time Asymmetry.* Cambridge, U.K.: Cambridge University Press, 1994.

Hardy, G. H. *A Mathematician's Apology.* Cambridge, U.K.: Cambridge University Press, 1993.

Heisenberg, Werner. *Physics and Beyond: Encounters and Conversations.* New York: Harper and Row, 1971.

Heisenberg, Werner. *Physics and Philosophy: The Revolution in Modern Science.* New York: Harper, 1958.

Holton, Gerald. *Einstein, History, and Other Passions: The Rebellion against Science at the End of the Twentieth Century.* Reading, Mass.: Addison-Wesley, 1996.

Home. D., and M.A.B. Whitaker. "Ensemble interpretations of quantum mechanics. A modern perspective." *Physics Reports* **210** (1992), pp. 223–317.

Jackson, J. D. *Classical Electrodynamics*. New York: John Wiley & Sons, 1975.

Jammer, Max. *The Philosophy of Quantum Mechanics—The Interpretations of Quantum Mechanics in Historical Perspective*. New York: John Wiley & Sons, 1974.

Jauch, Joseph M. *Foundations of Quantum Mechanics*, chapter 6. Reading, Mass.: Addison-Wesley, 1968.

Kolmogorov, A. N. *Foundations of the Theory of Probability*. New York: Chelsea, 1950 (first published in 1933).

Kolmogorov, A. N. "On the logical foundations of probability theory." In *Lecture Notes in Mathematics* **1021** (1983), pp. 1–5.

Kuhn, Thomas S. *The Structure of Scientific Revolutions*. Chicago: University of Chicago Press, 1970.

Lakatos, Imre. "Falsification and the methodology of scientific research programmes." In *Criticism and the Growth of Knowledge*. Imre Lakatos and Alan Musgrave, eds., pp. 91–195. Cambridge, U.K.: Cambridge University Press, 1970.

Laskar, Jacques. "Large-scale chaos and marginal stability in the solar system." In D. Iagolnitzer, ed., *Eleventh International Congress of Mathematical Physics*, pp. 75–120. Boston: International Press, 1995.

Laudan, Larry. *Beyond Positivism and Relativism: Theory, Method, and Evidence*. Boulder, Colo.: Westview Press, 1996.

Lichtenberg, Don B. *Unitary Symmetry and Elementary Particles*. New York: Academic Press, 1978.

Margenau, Henry. *The Nature of Physical Reality: A Philosophy of Modern Physics*. New York: McGraw-Hill, 1950.

Margurger, John H., III. "What is a photon?" *Physics Teacher* **34** (1996), pp. 482–486.

Martin-Löf, Per. "The definition of random sequences." *Information and Control* **9** (1966), pp. 606–619.

Martin-Löf, Per. "The literature on von Mises' collectivs revisited." *Teoria* **35** (1969), pp. 12–37.

Martin-Löf, Per. "Complexity oscillations in infinite binary sequences." *Zeitschrift für Wahrscheinlichkeitstheorie und verwandte Gebiete* **19** (1971), pp. 225–230.

Mermin, N. David. "Is the moon there when nobody looks? Reality and the quantum theory." *Physics Today*, April 1985, pp. 38–47.

Mermin, N. David. "Quantum mysteries refined." *American Journal Physics* **62** (1994), pp. 880–887.

Mills, Robert. "Gauge fields." *American Journal of Physics* **57** (1989), p. 493.

Mises, Richard von. *Probability, Statistics, and Truth*. New York: Macmillan, 1957.

Mises, Richard von. *Mathematical Theory of Probability and Statistics*. New York: Academic Press, 1964.

Nagel, Ernest. *The Structure of Science: Problems in the Logic of Scientific Explanation*. Indianapolis: Hackett, 1979.

Newton, Roger G. "Particles that travel faster than light?" *Science* **167** (20 March 1970), pp. 1569–1574.

Newton, Roger G. *Scattering Theory of Waves and Particles*. 2nd ed. New York: Springer-Verlag, 1982.

Newton, Roger G. *The Truth of Science: Physical Theories and Reality*. Cambridge, Mass.: Harvard University Press, 1997.

Omnès, Roland. *The Interpretation of Quantum Mechanics*. Princeton: Princeton University Press, 1994.

Plato, Jan von. *Creating Modern Probability: Its Mathematics, Physics, and Philosophy in Historical Perspective*. Cambridge, U.K.: Cambridge University Press, 1994.

Poincaré, Henri. *The Foundations of Science (Science and Hypothesis, The Value of Science, Science and Method)*. Lancaster, Pa.: Science Press, 1946. (Originally published in 1913.)

Popper, Karl R. *The Logic of Scientific Discovery*. New York: Basic Books, 1959.

Popper, Karl R. *Quantum Theory and the Schism in Physics*. Totowa, N.J.: Rowman and Littlefield, 1982.

Popper, Karl R. *Realism and the Aim of Science*, from the *Postscript to the Logic of Scientific Discovery*, W. W. Bartley, III, ed. Totowa, N.J.: Rowman and Littlefield, 1983.

Rosen, Joe. *A Symmetry Primer for Scientists*. New York: John Wiley & Sons, 1983.

Sakurai, J. J. *Modern Quantum Mechanics*. Menlo Park, Calif.: Benjamin/ Cummings, 1985.

Salmon, Wesley C. *Causality and Explanation*. Oxford: Oxford University Press, 1998.

Schulman, Lawrence S. *Time's Arrow and Quantum Measurement*. Cambridge, U.K.: Cambridge University Press, 1997.

Teller, Paul. *An Interpretive Introduction to Quantum Field Theory*. Princeton: Princeton University Press, 1995.

Ter Haar, D. *Elements of Thermostatistics*. New York: Holt, Rinehart and Winston, 1966.

Wannier, Gregory H. *Statistical Physics*. New York: John Wiley & Sons, 1966.

Webb, J. K., et al., "Search for time variation of the fine structure constant." *Physical Review Letters* **82** (1999), pp. 884–887.

Weinberg, Steven. *Dreams of a Final Theory*. New York, Pantheon, 1992.

190

Weinberg, Steven. *The Quantum Theory of Fields.* Cambridge, U.K.: Cambridge University Press, 1995.

Wheeler, John Archibald. *Geons, Black Holes and Quantum Foam.* New York: Norton, 1998.

Wigner, Eugene. "The unreasonable effectiveness of mathematics in the natural sciences." *Communications in Pure and Applied Mathematics* **13**, no. 1 (1960). Reprinted in *Symmetries and Reflection*, pp. 222–237. Bloomington: Indiana University Press, 1967.

Index

abelian, 115n
absorber theory, 161n
acausal, 125
accidental degeneracy, 118
acoustics, 28, 60
action at a distance, 8, 86, 88; spooky, 168
active tranformation, 112
alpha decay, 55
alpha-particle model, 39
analogies, 14
angular momentum, 114
anomalous magnetic moment, 97
anthropic principle, 17
approximation methods, 60
Aristotelian laws of motion, 31
Aristotelian universe, 43
Aristotle, 124
arrow of time: causal, 129, 150, 152, 159; cognitive, 151, 159; cosmological, 151, 161, 163; radiation, 159; thermodynamic, 151, 152, 155, 159, 162
Aspect, Alain, 173
associated production, 27
Auger effect, 49n
axial vector, 112

baryon, 99
Bayesian probabilities, 145
BCS theory of superconductivity, 28
Bell, John, 56, 169
Bell experiments, 170
Bell's inequality, 169
Bell's theorem, 169
Bernoulli, Daniel, 58
beta decay, 27, 28, 37, 55, 109
Bohm, David, 30, 52, 56, 136, 167
Bohr, Niels, 4, 37, 38, 52n, 167, 168, 177
Boltzmann, Ludwig, 29, 152
Bose-Einstein statistics, 91
Boyle, Robert, 3

Boyle's law, 8
bremsstrahlung, 96
Bridgman, Percy, 21
Broglie, Louis de, 34
Brouwer, Luitzen, 69
Brownian motion, 135
butterfly effect, 45

Caley-Klein parameters, 119
Cauchy, Augustin-Louis, 70
causal, 125; cycles, 164
causality, 25, 124, 159, 179; condition, 132; relativistic, 93, 129–132
cause, 124, 151; efficient, 124; final, 124; formal, 124; material, 124
Cepheid variables, 6
chaos, 45, 46n
circular polarization, 93
coarse graining, 47, 50, 136, 153, 155
coherence, 19, 53–56
collapse of the wave function, 140, 174
collective model, 38
Collins, C. B., 18
collision cross section, 100
complete set of dynamical variables, 48
Compton effect, 97
computers, 61
configuration space, 174
conservation laws, 114
constants of nature, 10, 19
constructive theories, 13
control, 127, 159, 164
conventionalism, 35
conventionalist stratagem, 37
Copenhagen interpretation, 137n
Copernicus, 35
cosmogony, 125n
cosmological constant, 32n, 37
Coulomb field, 120
Coulomb's law, 88
Cowan, Clyde, 38
creation operator, 92

193

197